Undergraduate Lecture Notes in Physics

Undergraduate Lecture Notes in Physics (ULNP) publishes authoritative texts covering topics throughout pure and applied physics. Each title in the series is suitable as a basis for undergraduate instruction, typically containing practice problems, worked examples, chapter summaries, and suggestions for further reading.

ULNP titles must provide at least one of the following:

- An exceptionally clear and concise treatment of a standard undergraduate subject.
- A solid undergraduate-level introduction to a graduate, advanced, or non-standard subject.
- A novel perspective or an unusual approach to teaching a subject.

ULNP especially encourages new, original, and idiosyncratic approaches to physics teaching at the undergraduate level.

The purpose of ULNP is to provide intriguing, absorbing books that will continue to be the reader's preferred reference throughout their academic career.

Series Editors

Neil Ashby
University of Colorado, Boulder, CO, USA

William Brantley
Department of Physics, Furman University, Greenville, SC, USA

Matthew Deady
Physics Program, Bard College, Annandale-on-Hudson, NY, USA

Michael Fowler
Department of Physics, University of Virginia, Charlottesville, VA, USA

Morten Hjorth-Jensen
Department of Physics, University of Oslo, Oslo, Norway

Michael Inglis
Department of Physical Sciences, SUNY Suffolk County Community College, Selden, NY, USA

More information about this series at http://www.springer.com/series/8917

Hassan Raza

Freshman Lectures on Nanotechnology

 Springer

Hassan Raza
Centre for Fundamental Research
Islamabad, Pakistan

ISSN 2192-4791   ISSN 2192-4805 (electronic)
Undergraduate Lecture Notes in Physics
ISBN 978-3-030-11731-3   ISBN 978-3-030-11733-7 (eBook)
https://doi.org/10.1007/978-3-030-11733-7

Library of Congress Control Number: 2018967441

This Springer imprint is published by the registered company Springer Nature Switzerland AG
The registered company address is: Gewerbestrasse 11, 6330 Cham, Switzerland

Dedicated to Tehseen, Ahmer, and Zuha.

Preface and Acknowledgements

Over the course of human civilization, we have witnessed various technological eras. These long and distant periods of history are remembered today based on the materials that were used, e.g. stone age, copper age, iron age and more recently silicon age. We are now witnessing a new technological age of nanotechnology, which would be remembered by how we use the materials by engineering them at the atomic and the molecular scale. This atomic and molecular engineering at the nanoscale distinguishes the nanotechnology era from the previous ones. Exploring this wonderful and exciting world of nanotechnology is what this book is all about.

The book covers the state-of-the-art knowledge about nanotechnology at first-year undergraduate level. High school students may also find the contents useful and interesting. My motivation in writing this book is to introduce the newcomers to what I would like to call the greatest intellectual journey human mind has ever ventured.

With this motivation, I offered a seminar course at the University of Iowa back in 2011. The class met for one hour each week for about 14 weeks. The first half of the semester was based on the lectures covering the content covered in this book. The second half of the course was exclusively based on the student contributions in the form of discussions and PowerPoint presentations. Throughout the semester, nanotechnology and nanoscience laboratory visits were offered to the students—who enthusiastically availed these opportunities. Overall, the course was very well-received and the associated findings have been reported as a research article. I am grateful to all the students who took this course and helped me improve the contents of the lecture notes, which served as the precursor for the book.

This book has been arranged into 11 chapters. Chapter 1 provides a gentle introduction to nanotechnology, its significance, various areas involved and the tools and methods one may use in this new science and technology. We also discuss applications both at the single device level and the system level. Chapter 2 introduces the concept of particles, waves and wave-particle duality. This chapter gives an overview of the historical development of the wave picture of matter. Chapter 3 introduces quantum mechanics and its role in understanding physical and chemical phenomenon at the nanoscale. Chapter 4 builds upon earlier chapters to teach the concept of bonding and how atoms arrange themselves to fabricate complex physical structures. We discuss the electronic-structure and interactions of various nanostructures in Chap. 5, whereas Chap. 6 deals with the electronic transport properties of nanoscale devices. Spintronics and photonic devices are covered in Chaps. 7 and 8, respectively. The nanoscale fabrication and characterization at the nanoscale are discussed in Chaps. 9 and 10, respectively. Finally, we discuss the safety, health, environmental and societal aspects of this novel technology in Chap. 11.

I would like to thank my family to which this book is dedicated to. Without their understanding, support and patience over the past 7 years, this book would have not been a reality. Furthermore, I would like to thank Dr. T. Z. Raza for improving the content of the book by extensive proof-reading. I also thank Mr. A. Siddique for proof-reading the final manuscript to improve readability and accessibility.

It is a pleasure to acknowledge useful discussions with Prof. David R. Andersen, a dear friend and a very valuable colleague. I would also like to acknowledge and appreciate the vision behind the Associateship Program at the Abdus Salam ICTP (International Centre for Theoretical Physics), Trieste, Italy. In this context, I am grateful to Dr. Sandro Scandolo and Dr. Ralph Gebauer for hosting my Associateship visit at the Abdus Salam ICTP that had been instrumental in bringing this project to completion. I would like to further thank Ms. Francesca Prelazzi of ICTP for facilitating my Trieste trip. This acknowledgment section would not be complete without remembering Dr. Abdus Salam.

I would like to thank Claus Ascheron, Adelheid Duhm and Elke Sauer at Springer Verlag for their collaboration and patience to bring this project to fruition. I would also like to thank Ravivarman Selvaraj, Sindhu Sundararajan and their production team at Springer Verlag for their commitment towards this book.

The initial version of the book was written at the University of Iowa, Iowa City, USA. Significant additions were made in the Abdus Salam ICTP, Trieste, Italy, and the final version was prepared in the Centre for Fundamental Research (CFR), Islamabad, Pakistan.

Islamabad, Pakistan   Hassan Raza

Contents

Chapter 1
Introduction

Nanotechnology is the art and science of designing materials, devices and systems at the nanoscale. Nanomaterials usually have critical dimension less than 100 nm, where one nanometer is a billionth of a meter, i.e. $1\,nm = 10^{-9}\,m$. To give an idea of the length scale, one may imagine that there are usually two to five atoms in one nanometer, depending on the type of atoms. Thus, nanotechnology boils down to working at the atomic scale ranging from a few atom systems to as many as millions or even billions of atoms packed into a cube of 100 nm side each.

The dimensionality is yet another important aspect in Nanotechnology as shown in Fig. 1.1.[1] Zero-dimensional structures are defined as the ones that have nanoscale features in all the three dimensions, e.g. organic molecules, quantum dots, nanocrystals, Bucky balls or C_{60}, etc. One-dimensional structures are the ones, which have nanoscale features in two dimensions and have substantial size in the remaining one dimension, e.g. carbon nanotubes, silicon nanowires, graphene nanoribbons, etc. Two-dimensional structures have nanoscale features in one dimension and are quite large in the remaining two dimensions, e.g. graphene, etc.

There is no concept of a three-dimensional nanostructure or nanomaterial, simply because in such a structure, no dimension is restrained in size. However, one may imagine fabricating or synthesizing a three-dimensional structure by using zero-dimensional, one-dimensional, or two-dimensional constituent nanomaterials stacked together. A C_{60} crystal in 3D is one such example, where C_{60} molecules are stacked in three dimensions. These nanomaterials are indeed very interesting and manifest novel properties due to this stacking.

---

[1] In a ball and stick model, ball represent atoms and the bonds between atoms are represented by sticks.

© Springer Nature Switzerland AG 2019

H. Raza, *Freshman Lectures on Nanotechnology*, Undergraduate Lecture Notes in Physics, https://doi.org/10.1007/978-3-030-11733-7_1

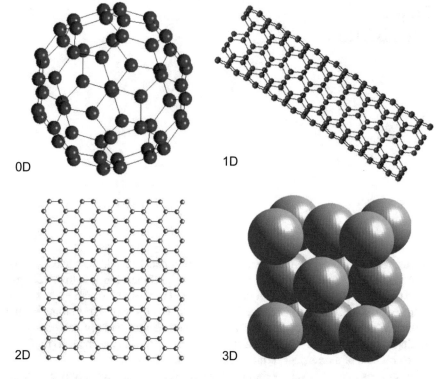

0D

1D

2D

3D

Fig. 1.1 Dimensionality illustrated by using ball and stick models. Atoms may be arranged in zero-dimensional (0D) nanostructures, e.g. a C_{60} molecule or Bucky ball. One-dimensional (1D) nanostructures include carbon nanotubes (CNT). Two-dimensional nanomaterials include graphene. Three-dimensional nanostructures have nanoscale constituents, e.g. C_{60} crystal, schematically displayed by spheres

1.1 Historical Perspective

Over the past few decades, the field of nanotechnology has attracted a lot of attention. We have already witnessed the impact in various industrial applications. *Microprocessors* do not have microscale devices anymore. The nanoscale transistors form the basis for these processors nowadays, and hence rightfully should be called *Nanoprocessors*. Integrated Circuit (IC) chips routinely have critical dimensions less than 100 nm. It is not uncommon to find features as small as 7 nm in commercial products[2] and in fact, less than 1 nm in research grade emerging devices.

Nanotechnology is precisely the driver behind the ever-increasing information processing power of our computers. It is helping us find solutions for world's energy and clean water problems. Moreover, there are plenty of examples using nanomate-

---

[2] As of 2018.

rials to improve performance, while keeping the cost down in medicine, pharmaceuticals, chemicals, and materials manufacturing.

In this context, physicists should be accoladed for almost a century old concept of atomic-scale picture of material and its detailed understanding for various physical phenomenon. Chemists should be credited for almost a century or more of engineering, synthesis, manipulating, and probing of molecules, which usually are of sub-nanometer size. In biology and medicine, associating the functionality of nanoscale DNA molecules with the mysteries of life has been quite rewarding. For engineers, with the progress in physical and biological sciences, the use of these amazing and exceptional nanomaterials for novel applications has always been a goal.

As a civilization, our use of nanomaterials goes well beyond the past century. The classical example is the use of metal nanoparticles to synthesize colored glass a few centuries ago, which have been widely used in church windows all over the Europe. In fact, the colored glass has not faded over centuries, and has provided a good template to design glass for space exploration, where radiation damage is a major concern.

More recently, graphene is a single atom thick nanomembrane extracted from graphite, the material used in *lead* pencils. This material has caught huge attention due to its excellent electrical, mechanical, and optical properties since 2004.

Although the nano *gold rush* is quite recent, however, the most modern realization is attributed to the American physicist Feynman, who pointed towards the possibility of atomic scale engineering in a 1959 talk at the American Physical Society (APS) meeting at Caltech (California Institute of Technology) with the title *there is plenty of room at the bottom*[3] and a meaningful comment that,

In the year 2000, when they look back at this age, they will wonder why it was not until the year 1960 that anybody began seriously to move in this direction.

—Richard P. Feynman

He started his talk with,

What I want to talk about is the problem of manipulating and controlling things on a small scale.

—Richard P. Feynman

And asked interesting questions like,

Why cannot we write the entire 24 volumes of the Encyclopedia Brittanica on the head of a pin?

—Richard P. Feynman

---

[3]Feynman, R.P. Engineering and Science 22–36 (February 1960).

1.2   Bottom-Up Approach

Most of our technological advances have been centered around what is usually called the *top-down* approach. Consider a simple example of a chair. We chop down a tree, process the wood or simply cut it in blocks, which may be glued together to build a chair. This top-down approach is even more evident in technologies centered around the principle of miniaturization. Consider integrated circuits for example, where we deposit films on large silicon wafers, pattern them by using lithography, and chemical or physical etching to fabricate smaller structures.

This top-down approach, although very successful, has fundamental bottlenecks when approaching the atomic scale. These shortcomings indeed come from our inability to control atoms at individual level in a collection of say million atoms. Familiar problem of edge roughness is precisely the result of this bottleneck. In other words, one may not prepare an extremely smooth surface by using the top-down approach. On the contrary, if one observes nature including life, it is the *bottom-up* approach that nature follows. The holly grail in this area is to truly understand the marvels of nature and to replicate them artificially in a laboratory.

In this context, the lessons learned from nature in its physical, chemical and biological functions has led to a new way of thinking, called the bottom-up approach. Instead of starting big, one may start small, e.g. from atoms and put them together in a desired configuration dictated by the laws of nature, albeit with some manipulation. An elegant example is the self-assembly of molecules on various substrates, where one dips the sample in a molecular solution and the molecules start arranging themselves on the surface in a perfect order. In the top-down approach, such arrangements would have been extremely tedious, if not impossible.

The unified bottom-up approach towards material, device, and system design is quite revolutionary, not only from the point of view of immense technological possibilities, but also from commencing a new way of thinking. The properties of individual atoms at the nanoscale control the microscopic and even macroscopic properties. This fact makes the understanding of the material properties at the nanoscale even more compelling. Furthermore, by using the bottom-up approach, nature performs various tasks in an environment-friendly way. The hope is that by learning from nature and by using the bottom-up technology, we may reduce our impact on the environment as a global community.

1.3   Why Nano is Different?

To answer the question that *why nano is different* both scientifically and technologically, one has to understand how matter behaves differently at the nanoscale. Classical mechanics, also known as Newtonian mechanics, has been widely used for every day phenomenon at the macro scale and even in some situations at the microscale. However, at the nanoscale when atomic details become important, the experimental observations may not be explained by the laws of classical mechanics. For example, according to classical mechanics, energy inside a system may have continuous values. However, if one looks at the light emitted or absorbed by the atoms, it has discrete wavelengths or frequencies instead of a continuous emission or absorption spectrum as predicted by the laws of classical mechanics.

To explain this quantized nature of energy exchange, a new kind of science had to be invented, which is now known as Quantum Mechanics. The physical phenomenon at this scale is covered under the umbrella of Quantum Physics, and the chemical processes are handled within the discipline of Quantum Chemistry. The understanding brought by the Quantum revolution in the early part of the twentieth century is precisely what one needs to understand how Nanotechnology works. Basic questions like, *why the color of a gold ring becomes red if the diameter hypothetically becomes on the order of 10 nm? or why the roses and tulips have these beautiful colors?* have answers hidden in understanding nanotechnology.

One of the properties of matter at the nanoscale is the quantization of energy levels. Yet another is the wave nature of matter, which is not only intriguing but often times mind boggling. We are very familiar with the particle nature of matter, where electrons behave as balls, exchanging energy and momentum in collisions and even colliding off various barriers. On the other hand, the concepts like propagation, transmission, reflection and interference are associated with waves. It is difficult to imagine matter acting as waves at macro or micro scale. If we take the example of electrons, they have the ability to propagate like light, partially getting transmitted through or reflected from surfaces and even causing constructive and destructive interference. The most challenging aspect is to imagine electrons having a wavelength and a frequency just like waves. Thus, sometimes matter behaves like waves and sometimes it behaves like good old particles - giving rise to the principle of *Wave Particle Duality*, i.e. matter may act either like a wave or a particle, but not both at the same time.

For matter waves, if an observer makes a measurement in a finite time or a finite space, it leads to an uncertainty in the outcome of the measurement. This is called the *Uncertainty Principle* that relates uncertainty in two pairs, i.e. (energy, time) and (momentum, space), which we discuss later.

In conclusion, due to the fact that two different kinds of mechanics, i.e. Quantum versus Classical, govern dynamics at the nanoscale and the macro/micro scale, respectively, the properties of material at the nanoscale are drastically different from the everyday macro or micro scale. To probe and engineer these properties, one needs new capabilities and equipment. At the macro scale, one may use the *naked eye* to look at the devices, but at the microscale, an optical microscope is required. However, to *see* at the nanoscale, one has to use electron microscopes, where electron waves instead of light waves are used.

1.4   Nanomaterials

While it is impossible to discuss an exhaustive list of nanomaterials, we focus on few representative nanomaterials in this section. An enthusiastic reader is encouraged to explore further based on interest.

Let us begin the discussion with 0D nanoparticles as shown in Fig. 1.2a. Such nanoparticles may be fabricated by using chemical, physical, or even biological means. They have found applications in diverse areas like electronics, sports equipment, cosmetics, etc.

The next on the list are 1D nanowires as shown in Fig. 1.2b, which have excellent electrical characteristics and may be fabricated by using bottom-up methods or top-down techniques like lithography. Nanofibers shown in Fig. 1.2c have also attracted huge attention due to their excellent mechanical as well as electrical characteristics. Furthermore, the nanofilms have remarkable electrical and optical properties. These are widely used for making protective coatings as filters. In addition, they are used in

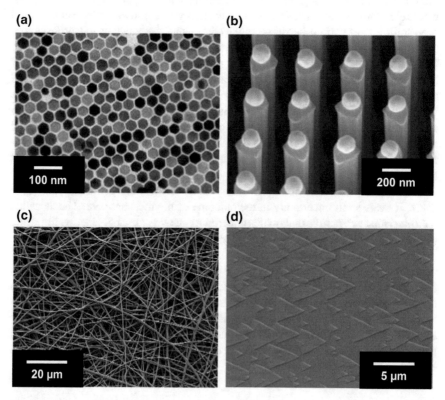

Fig. 1.2 Nanomaterials. **a** CuInSe$_2$ Nanoparticles, **b** GaAs Nanowires. Au nanoparticle catalyst is visible at the top. **c** Poly(vinyl alcohol) (PVA) Nanofibers. **d** GaP nanofilm with faceted morphology [1–4]

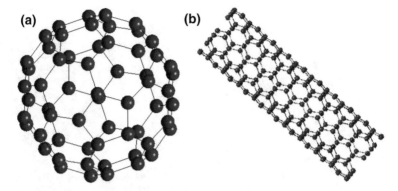

Fig. 1.3 Carbon Nanomaterials. **a** Fullerene (C_{60}) molecule, and **b** Carbon Nanotube

the fabrication of IC chips and electronic components. One such nanofilm is shown in Fig. 1.2d.

Carbon nanomaterials have also attracted considerable attention in the past few decades due to their excellent physical, chemical, and biological properties. Fullerene (Buckyball, C_{60}) molecules are the example of 0D carbon nanomaterials as shown in Fig. 1.3a. In fact, one may encapsulate other atoms inside the Buckyball cage. By incorporating magnetic atoms like Fe and Co, one obtains single molecule magnets as discussed in Chap. 7.

Carbon nanotubes (CNTs) are 1D structures of rolled benzene rings as shown in Fig. 1.3b. CNTs are known to have unique electrical and optical properties, which depend on their diameter and chirality, i.e. how they are rolled, resulting in peculiar atomic arrangement at the edges. Various atoms may also be encapsulated inside CNTs. While a single wall CNT is shown in Fig. 1.3b, multiwall CNTs exist as well. In addition to the electrical and photonic devices, CNTs have also been explored for their mechanical properties.

Graphene is a 2D monolayer of carbon atoms with hexagonal arrangement as shown in Fig. 1.4a. Due to its linear dispersion in the electronic spectrum, it has unique electrical and optical properties. Furthermore, two graphene layers may be stacked on top of each other to form a bilayer graphene membrane as shown in Fig. 1.4b.

Structures like porous anodic alumina (PAA) are also quite interesting as shown in Fig. 1.5. These pores may be used to store various materials at the nanoscale or may act as templates for nanomaterial growth and synthesis. By using nanofabrication techniques like atomic layer deposition (ALD), dielectric films may be deposited inside the pores as shown in Fig. 1.5.

There is on-going research to improve existing materials by adding the nanomaterials to their composition. Conventionally used building materials like steel and cement are under exploration. In nanocement, the conventional cement is reinforced with the nanomaterials to achieve better compressive strength.

(a) **(b)**

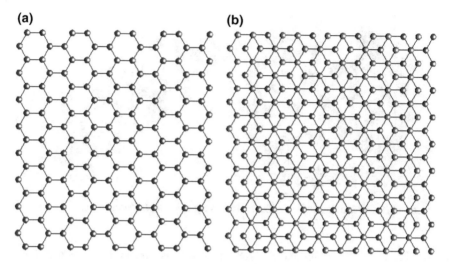

Fig. 1.4 Graphene. **a** Monolayer, and **b** Bilayer

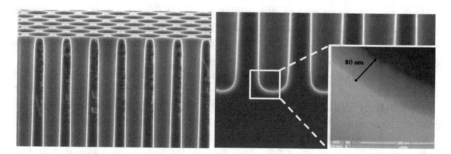

Fig. 1.5 Porous anodic alumina (PAA) with 80 nm dielectric deposited inside the pores (courtesy of Oxford Instruments)

Biomaterials are a new avenue where nanotechnology and nanomaterials have made their mark as well. An artificial retina is shown in Fig. 1.6. The hope is that one day, such retinas would replace the damaged retinas to restore vision. Nanotechnology is also expected to play a major role in fabricating artificial organs to improve quality of life of millions of people world-wide.

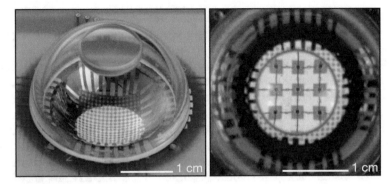

Fig. 1.6 Artificial retina [5]

1.5 Applications

Nanotechnology is expected to pave way for high-performance and low-power computing devices, like computers, laptops, tablets, cell phones, etc. The use of nanotechnology in agriculture is projected to result in better yield and more efficient use of resources. Nanomedicine is also an important area that may provide medical choices engineered at the nanoscale, e.g. biochips, bio-nanorobots, etc. Drug delivery at the nanoscale may provide affordable choices with minimal side effects. High-performance, light-weight and green materials are expected to provide diverse design options for engineers and professionals. Affordable choices in defense sector might include better sensors, improved bullet and explosion proof materials, nano armor, smart systems, etc. Consumers are expected to enjoy various clean energy choices based on nanoengineering of materials and devices, e.g. solar cells, thermoelectric, solar-thermal, etc. Environmental sensing and monitoring choices are expected to play an important role in improving the quality of life. Finally, the bottom-up approach towards nanotechnology may result in unprecedented choices in manufacturing that would result in defect-free materials and devices at the atomic scale.

Apart from this, nanotechnology truly has a great promise to provide us with clean energy by designing and engineering various technologies at the atomic scale. High-performance and low-cost solar cells as well as thermoelectric devices and solar-thermal applications seem to provide a promising solution. Furthermore, the use of nanotechnologies is expected to result in Green Manufacturing. The bulk of existing nano-manufacturing is based on the material Silicon, whose processing is highly inefficient, energy consuming and simply wasteful. A significant amount of green house gas emission is unavoidable during silicon processing due to the use of Fluorine-based chemistry, which should be kept under check according to the various environmental protection protocols. There is a new area of Graphene Nanoelectronics,[4] which is expected to result in significant reduction in the green house emission

---

[4]Raza, Hassan (Editor), Graphene Nanoelectronics: Metrology, Synthesis, Properties and Applications, (Springer, Heidelberg, 2012).

due to oxygen based etching chemistry. Apart from this, with the possibility of the use of self-assembly in graphene synthesis, significant waste reduction is expected in this emerging technology.

The novel approach of nanotechnology based on the atomic scale engineering is expected to result in wide ranging applications in computation, agriculture, medicine, drugs, materials, defense, clean energy, environment, manufacturing, travel, and sports. Nanotechnology has not only been instrumental in improving the existing products in these areas, but is prospective for novel applications inconceivable at this time. Consider nanorobots that one day may enter the body through an injection, track the cancer cells and kill them at will, all within the span of minutes! The use of nanotechnology in engineering materials has ushered a new era of high-performance and light-weight nanomaterials, which has revolutionized the industry from multi-million dollar aviation products to high-end sports equipment. The everyday applications include sunscreens (containing nanoparticles), tennis balls and tennis racquets, stain-free clothing and mattresses, polymer films used in displays for laptops and cell phones, coatings for easier cleaning glass, bumpers and catalytic converters, protective and glare-reducing coatings for eyeglasses, to name a few. The use of nanomaterials in manufacturing may further lead to novel materials, e.g. high performance composite alloys, durable plastics, functional clothing, etc.

The bottom-up approach towards Nanotechnology following the Nature's principles opens up the possibility of sustainable technology development. In this section, we highlight various areas, which may directly benefit from nanotechnology and provide a few notable examples for each discipline.

The field of electronics has already been influenced by Nanotechnology, giving rise to the area of Nanoelectronics. Nanoscale transistors are the building blocks of today's integrated circuits. Historically, transistors have been scaled according to the top-down approach, enabling the packing of more transistors per unit area, and resulting in higher performance and reduced unit cost - a trend emphasized in Moore's law. With the continued scaling, we have already crossed the 100 nm limit in early 2000. As of today,[5] 7 nm node chips are in production and 5 nm node chips are on their way into commercial integrated circuits. It is estimated that by using the top-down approach, we may be able to scale down to about 5 nm and yet make efficient integrated circuits. Irrespective of the debate about the lower limit, this trend has to come to an end, since we are approaching the atomic limits. To give an idea, there are only about 10 Silicon atoms in 5 nm channel length on a Silicon surface! Enormous amount of research is being carried out to extend the physical limits beyond what is possible with the top-down approach. By using the bottom-up approach and novel physical and chemical principles at the nanoscale, the researchers may be able to push the envelope of scaling. The role of bottom-up approach towards Nanotechnology to solve this problem is yet to be seen.

---

[5] 2018.

Mechanical motion at the nanoscale has always fascinated scientists and engineers. To make use of this mechanical flexibility, one may engineer Nano-electromechanical Systems (NEMS) based on nanomaterials. Compared to their macro and micro[6] counter parts, NEMS are expected to have better performance and bring new functionalities.

Nanoscale sensors have revolutionized the field of physical, chemical and biosensing. The primary advantage nanosensors have is the increased surface to volume ratio, compared to the micro or macro sensors, where most of the atoms are at the surface in a nanomaterial. Consider carbon nanotube or C_{60} molecule, where virtually all the atoms are on the surface. This simple fact leads to increased sensitivity and detection levels, which were not possible before.

Electrons have an inherent property called spin, which leads to magnetism. Spin may be in two different orientations, usually labeled as up-spin ($\uparrow$-spin) or down-spin ($\downarrow$-spin). This bistable configuration leads to computing or memory applications, similar to ON and OFF states of a transistor. Devices based on the spin of an electron form an active area of research, development and commercial activity, called Spintronics. Hard disks are such devices, where information is stored in the magnetic orientation of one bit, which is precisely analyzed later by detecting the orientation.

Nanomaterials have also revolutionized photonics. Consider LASER (light amplification by stimulated emission of radiation) devices and LEDs (light emitting diodes), which have become an integral part of our everyday life. Proper functioning of these high performance photonic devices depends on the precise control and preparation of nanoscale films and surfaces. LEDs continue to be an important class of devices due to their extended lifespans and high conversion efficiencies. The reader may have noticed the use of LEDs in traffic and automobile lights. One should expect to witness rising popularity of the LED fixtures in residential and commercial buildings. In this context, one may have to not worry about changing the lighting fixtures for decades according to the current state of the art. We may very well be able to improve the technology to the point that one would not have to change a light fixture in a life-time! Furthermore, one may envision having flexible displays, which can be rolled in a bag for transportation or when not in use!

Energy harvesting is an important subject of our time. Clean and renewable energy is one of the key challenges that we face today. The problem is that in case of other energy sources like nuclear, thermal, geothermal, fossil, hydel, etc, the source is at the surface of the earth, which leads to negative consequences for the environment and compromises sustainable development. In the case of solar energy, the source is clearly far away and therefore harvesting this energy does not result in any environmental foot-print directly on the earth, apart from the environmental impact during the material synthesis and packaging, and of course during disposal at the end of the life-cycle.

Nanomaterials have played an important role in solar energy harvesting, and it is anticipated that these will continue to play a key role in future technology development by enhancing the conversion efficiencies. Waste energy harvesting is also

---

[6]Micro-electromechanical Systems (MEMS).

an important avenue, where the energy may be converted into electricity by using thermoelectric devices. Although they have come a long way, but still important breakthroughs need to be made for their widespread applications. For the energy storage applications, nanomaterials have already shown to improve battery performance, and have even led to the development of supercapacitors.

Yet another area is Nanomedicine, the application of Nanotechnology to medicine. Targeted drug delivery is an example, where the objective is to engineer an autonomous or semi-autonomous vehicle carrying the drug to a specific location inside the body. From this description, robots sent to Mars come to mind, albeit at a different scale altogether. In this analogy, one may envision molecular engineered robots injected in blood stream to automatically detect and kill cells by physical (radiation) or chemical (drug) means locally, and not to expose the healthy cells to severe conditions. Such a technology would truly improve the medical practice one day. Along similar lines, interdisciplinary areas like Nanobiotechnology have led to the development of systems like lab on a chip. Another set of applications include detecting pathogens inside food at chip level for food safety and security.

In this section, we have tried to outline a few examples and directions, but the applications are truly endless. In short, it is simply not possible to predict the impact that Nanotechnology would have in our society, but surely *there is plenty of room!*

Problems

1.1 What is the number of transistors in the latest Intel chip?
1.2 What is the number of transistors in the latest flash memory chip? You may pick any manufacturer.
1.3 What is the transistor size (critical dimension) in the latest Intel chip?
1.4 What is the transistor size (critical dimension) in the latest flash memory chip? You may pick any manufacturer.
1.5 Name a 0D nanomaterial not mentioned in this chapter.
1.6 Name a 1D nanomaterial not mentioned in this chapter.
1.7 Name a 2D nanomaterial not mentioned in this chapter.
1.8 Briefly explain the top-down approach towards nanotechnology. Give one example.
1.9 Briefly explain the bottom-up approach towards nanotechnology. Give one example.

Research Assignment

R1.1 Manufacturing is an important aspect of Nanotechnology. Various examples include the use of molecular robots, nanocement, nanomaterials, nanopaints, sustainable manufacturing, etc. Pick a topic of your choice about how nanotechnology is affecting manufacturing in contemporary society, and write a one-page summary.

References

1. Aldakov et al., J. Mater. Chem. C **1**, 3756 (2013)
2. Åberg et al., IEEE J. Photovolt. **6**, 185 (2016)
3. Oktay et al., Conference Series: Materials Science and Engineering **64**, 012011 (2014)
4. Warren et al., IEEE 42nd Photovoltaic Specialist Conference, pp. 1–4 (2015)
5. Ko et al., Nature **454**, 748 (2008)

Chapter 2
Particle, Waves, and Duality

In a basic physics course, one is introduced to the concept of particles and waves. The understanding evolves with the intuitive concepts of particles colliding with each other to exchange energy and momentum; and waves interacting with each other to create interference patterns. Such everyday observations may be explained by using Classical mechanics, also known as Newtonian mechanics. However, at the atomic scale, this simple everyday intuitive picture does not hold. Whether an electron behaves as a particle or a wave depends on the circumstances. Sometimes, it may collide with other electrons and barriers just like a particle. At times, it behaves like waves and in fact, creates interference patterns! It may also transmit through barriers, a phenomenon called quantum mechanical tunneling, which has no classical analog.

This dual nature of matter at the atomic scale is referred to as the *wave-particle duality*. The confinement of these matter waves in a region leads to the quantization of energy and momentum values - similar to the quantization of nodes in a guitar string to generate different tones. To understand this behavior at the atomic scale, a new kind of mechanics was developed in the early part of the twentieth century, now known as the quantum mechanics. From the name, it is evident that it is a kind of mechanics that results in quantized values. In contrast, the classical or Newtonian mechanics is also referred to as the continuum mechanics due to the non-quantized nature of various quantities. In this chapter, we study the behavior of matter at the nanoscale, starting from an introduction to particles and waves. This discussion leads us to the next chapter about the atomic picture of matter.

© Springer Nature Switzerland AG 2019

H. Raza, *Freshman Lectures on Nanotechnology*, Undergraduate Lecture Notes in Physics,
https://doi.org/10.1007/978-3-030-11733-7_2

Fig. 2.1 Particle Dynamics.
a Rectilinear motion is
described by the speed of the
particle and its mass.
b Circular motion is
represented by the radius (r)
of the circular trajectory, the
speed and the mass. **c** Two
charged particles, q_1 and q_2,
separated by a distance R
experience electrostatic force
(F) given by the Coulomb's
law

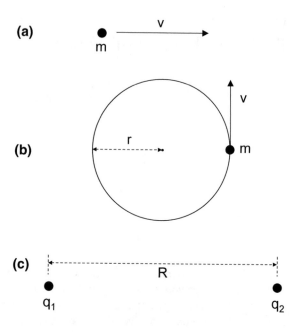

2.1 Particles: Classical Mechanics

The concept of a particle as an object with a mass m, either at rest or moving with a
speed[1] v along a straight line as shown in Fig. 2.1a, is mathematically expressed as:

$$v = \frac{dx}{dt} \tag{2.1}$$

which has the units of [m/s].

According to the Newton's first law, such an object maintains its speed and the
direction of travel in the absence of a force. If this is not the case and the velocity is
not constant, one may define acceleration[2]; defined as,

$$a = \frac{dv}{dt} \tag{2.2}$$

which has the units of [m/s²]. The product of the acceleration and the mass equals
force (F). According to Newton's second law, it is defined as,

---

[1] While speed is a scalar quantity, velocity is a vector. Scalar quantities are represented by magnitude
only, whereas vectors have both magnitude and direction. In this text, we would not worry about
the vector representation of these quantities, and would work with the magnitude only.

[2] Acceleration is a vector quantity. Even a change in the direction with constant speed gives rise to
acceleration.

$$F = ma = m\frac{dv}{dt} = \frac{d(mv)}{dt} \tag{2.3}$$

which has the units of [N].

Newton's third law ensures conservation of momentum and energy by stating that for every action, there is an equal and opposite reaction. In other words, energy cannot be created or destroyed, but it may change from one form to another. Mathematically, one may use this conversation of momentum and energy to derive various relationships.

For linear motion, one may define the linear momentum as follows:

$$p = mv \tag{2.4}$$

which has the units of [kgm/s].

Comparing (2.3) and (2.4), one may rewrite Newton's second law as:

$$F = \frac{dp}{dt} \tag{2.5}$$

For a particle with an angular motion around a circle with radius r as shown in Fig. 2.1b,[3] the angular momentum is defined as:

$$L = mvr \tag{2.6}$$

which has the units of [kgm^2/s]. The centripetal force, which keeps it in a circle, is defined as:

$$F_c = \frac{mv^2}{r} \tag{2.7}$$

There are two kinds of energies involved in any physical process, kinetic energy (E_K) related to the motion of the particle and potential energy (E_P) assigned to a particle depending on its location and position in a field. Kinetic energy is defined as follows:

$$E_K = \frac{1}{2}mv^2 = \frac{p^2}{2m} \tag{2.8}$$

whereas, the potential energy is field dependent. For example, in gravitational field, it is given by:

$$U_g = mgh \tag{2.9}$$

where h is the height from the surface of earth and g is the gravitational acceleration. Energy has the units of Joules (J), where [J $= $ kgm^2/s^2].

For our purpose, the most important field is the electrostatic field due to the charged nature of particles. The electrostatic force is given by Coulomb's law as follows:

---

[3] Circular motion is a special type of curvilinear motion.

$$F_e = K \frac{q_1 q_2}{R^2} \tag{2.10}$$

and the potential energy is given as:

$$U_e = K \frac{q_1 q_2}{R} \tag{2.11}$$

where $q_{1,2}$ are the two charges separated by the distance R as shown in Fig. 2.1c. K is the proportionality constant, and has a value of 9×10^9 m/F in free space.

For the electrostatic field, the total energy is given as:

$$E_T^e = E_K + U_e = \frac{1}{2} m v^2 + K \frac{q_1 q_2}{R} = \frac{p^2}{2m} + K \frac{q_1 q_2}{R} \tag{2.12}$$

Consider a hydrogen atom, which consists of one electron orbiting the nucleus with charge $q_1 = -q$, where $q = 1.6 \times 10^{-19}$ C, and one proton in the nucleus with charge $q_2 = +q$. For the electron with mass $m_e = 9.11 \times 10^{-31}$ kg, the total energy is,

$$E_T^e = \frac{1}{2} m_e v^2 - K \frac{q^2}{R} = \frac{p^2}{2m_e} - K \frac{q^2}{R} \tag{2.13}$$

This is an important equation, which we use in the next chapter while discussing the hydrogen atom in the context of the atomic picture of matter. Similarly, the electrostatic force given by (2.10) becomes,

$$F_e = -K \frac{q^2}{R^2} \tag{2.14}$$

2.2 Waves

Waves may be represented by an amplitude varying as a function of space and time. Consider Fig. 2.2, where a wave is shown as a function of time (t) as well as real space (x). In the time domain, one may define the time period T, for which a wave completes one cycle as shown in Fig. 2.2a. Corresponding to this time period, the frequency is defined as:

$$f = \frac{1}{T} \tag{2.15}$$

with the units of cycles per second or *Hertz* (abbreviated as Hz). One should note that one cycle equals 360° - similar to going around a circle to complete one circular cycle. Another unit of measuring angle is called *radian* (abbreviated as rad) with the following definition,

$$2\pi \, \text{rad} = 360° \tag{2.16}$$

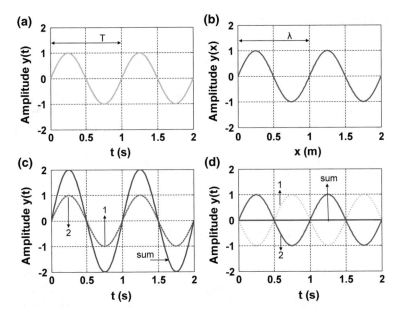

Fig. 2.2 Wave Phenomenon. **a** Time Period. **b** Wavelength. **c** Constructive Interference for zero phase. **d** Destructive Interference for 180° phase difference

or in other words, one cycle has 2π *radians*,[4] which gives rise to a new quantity, called angular frequency, given as:

$$w = 2\pi f = \frac{2\pi}{T} \qquad (2.17)$$

and has the units of [rad/s].

Similarly, in the real space, the length over which a wave completes one cycle is called the wavelength λ as shown in Fig. 2.2b. In analogy with the time-domain analysis, one may define wave number (k) as:

$$k = \frac{2\pi}{\lambda} \qquad (2.18)$$

which has the units of [rad/m].

Finally, the speed of wave propagation (v) is given as:

$$v = f\lambda \qquad (2.19)$$

When two waves interact with each other, there could be some time difference ΔT between the corresponding maxima or minima. This time difference may be

---

[4]Both degrees and radians are dimensionless units.

translated to a phase difference (ϕ) in terms of an angle as follows:

$$\phi = \frac{\Delta T}{T} 2\pi \text{ rad} = \frac{\Delta T}{T} 360°$$

(2.20)

For a time difference of one cycle, the phase difference is 2π rad or 360°, whereas a time difference of half a cycle gives a phase difference of π rad or 180°. Two waves with no phase difference are shown in Fig. 2.2c. Such waves overlap with each other and thus give rise to constructive interference. The two waves shown in Fig. 2.2d have a phase difference of half a cycle, i.e. π rad or 180°. Such waves completely cancel each other, resulting in destructive interference.

2.3 Photon: Elementary Particle of Light

For a long time, light was thought to be a wave. In the early part of the twentieth century, while explaining the black body radiation phenomenon, Max Planck proposed that light consists of elementary particles, which we now call photons. For each photon, the energy (E) is given as:

$$E = hf$$

(2.21)

where f is the frequency of the photon and h is a universal constant, called Planck's constant, and is given as:

$$h = 6.62 \times 10^{-34} \text{ Js}$$

(2.22)

In terms of the angular frequency, the energy (E) is given as:

$$E = \hbar w$$

(2.23)

where $\hbar$ is called the reduced Planck's constant and given as,

$$\hbar = \frac{h}{2\pi}$$

(2.24)

For n number of photons of angular frequency w, the energy is given as[5]:

$$E_n = n\hbar\omega$$

(2.25)

where $n = 1, 2, 3, \ldots$

---

[5]Photons are Bosons. Therefore, multiple bosons may occupy the same energy state. In contrast, electrons are Fermions, and one energy state may house only one electron.

Assigning photon as an elementary particle of light has important conceptual consequences. This means that energy of photon is quantized and may only be exchanged in discrete quantities depending on the frequency of the light! This led to the concept of quanta of light and was very effectively used by Einstein to explain the photoelectric effect. With ultra-sensitive single photon detectors accessible to researchers these days, it is possible to detect single photons routinely in the laboratory. At a given energy or frequency, the number of photons available are given by Bose–Einstein statistics as:

$$n(\omega) = \frac{1}{e^{\hbar\omega/k_B T} - 1} \tag{2.26}$$

where k_B is the Boltzmann's constant and T is the temperature in K.

Consequently, it turns out that the linear momentum (p) carried by a wave quanta is given as:

$$p = \hbar k \tag{2.27}$$

By using (2.18), the above equation becomes,

$$p = \frac{2\pi\hbar}{\lambda} = \frac{h}{\lambda} = \frac{hf}{v} \tag{2.28}$$

where v is the speed and equals $c = 3 \times 10^8$ m/s for free space propagation. This implies that the momentum increases with the decreasing wavelength or increasing frequency and vice versa.

According to (2.21), the energy depends on the frequency of vibration alone, since the Planck's constant is a universal constant. The energy should be conserved in the absence of any energy dissipation processes, e.g. heat generation, while the frequency remains constant and depends only on the generation process. We further discuss this in the next chapter. On the other hand, the velocity and hence the momentum changes in different media. The speed of light in glass is smaller than that in vacuum for example. Hence, for the same frequency of vibration, the wavelength changes with varying speed and vice versa with varying media of propagation (2.19).

2.4 Matter Waves

In the previous section, we talked about photons as elementary particles and yet described them by using frequency and wavelength - concepts which are attributed to treating light as a wave. On the other hand, one may imagine a matter particle acting as a wave. In which case, we may call it a matter wave and still calculate its properties by using a mass. Such matter waves are localized in a region of space and are in fact called wave-packets as shown in Fig. 2.3. These wavepackets mathematically represent the probability of finding a particle, further discussed in the next chapter. Such

Fig. 2.3 Wave-packet representation of a particle. The particle has both the wave character as well as spatial localization

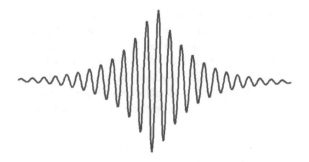

wave-packets have oscillatory behavior and are short ranged to fulfill the physical requirement of the matter to be localized.

In order to have a common ground between particles and waves, one may equate (2.4) and (2.28), thereby resulting in de Broglie equation as follows:

$$mv = \frac{h}{\lambda} \tag{2.29}$$

Rearranging in terms of the wavelength,

$$\lambda = \frac{h}{mv} \tag{2.30}$$

where λ is the wavelength of the matter wave, also known as the de Broglie wavelength. According to the equation above, increasing the velocity results in decreasing wavelength. As abstract as it may sound, these effects are routinely observed nowadays in laboratories all over the world.

The behavior of matter at the atomic and the nanoscale is quite peculiar as compared to the common day experiences and most of the times defies common sense. This fundamentally different regime is precisely the reason why nanotechnology is so interesting and carries immense possibilities of innovation. Since its conception over a century ago, we have only taped the tip of what we would like to call a *Quantum Iceberg*. The best stuff is yet to come . . .

Problems

2.1 For $E = 1\,\text{eV}$, calculate the frequency for a quantum particle. You may assume speed equal to that of the speed of light (c).

2.2 For $E = 1\,\text{eV}$, calculate the wavelength for a quantum particle. You may assume speed equal to that of the speed of light (c).

2.3 For $E = 1\,\text{eV}$, calculate the wavenumber for a quantum particle. You may assume speed equal to that of the speed of light (c).

2.4 For $E = 1\,\mathrm{eV}$, calculate the momentum. You may assume speed equal to that of the speed of light (c).

2.5 For a visible photon of wavelength $0.5\,\mu\mathrm{m}$, calculate the frequency and energy. What would be the color of light?

2.6 Calculate n (the number of Bosons) at $T = 300\,\mathrm{K}$ for the data in Problem 2.5.

2.7 For an electron traveling at $v = 1\,\mathrm{m/s}$, calculate its wavelength.

2.8 For an electron traveling at $v = 10^3\,\mathrm{m/s}$, calculate its wavelength.

2.9 For an electron traveling at $v = 10^6\,\mathrm{m/s}$, calculate its wavelength.

2.10 For an electron traveling at $v = 0.9\,c$, calculate its wavelength. c is the speed of light.

Research Assignment

R2.1 Various industrial and consumer applications of nanotechnology include plastics, textile fabrics, cosmetics, sports equipment, surfaces and coatings, etc. Pick a topic of your choice about how nanotechnology is impacting industrial and consumer applications, and write a one-page summary.

Chapter 3
Atomic Matter

If, in some cataclysm, all scientific knowledge were to be destroyed, and only one sentence passed on to the next generation of creatures, what statement would contain the most information in the fewest words? I believe it is the atomic hypothesis (or atomic fact, or whatever you wish to call it) that all things are made of atoms - little particles that move around in perpetual motion, attracting each other when they are a little distance apart, but repelling upon being squeezed into one another. In that one sentence you will see an enormous amount of information about the world, if just a little imagination and thinking are applied.

—Richard P. Feynman

Yet again, I quote from Feynman, emphasizing the importance of the atomic picture of matter. The word atom literally means *uncuttable* or *indivisible* in the Greek language and the concept that the matter consists of these *uncuttable* or *indivisible* building blocks was introduced in ancient Greece, of course without any experimental proof, but rather based on the physical intuition and imagination. We now indeed know that atoms are not indivisible. They consist of subatomic particles with a central region called the nucleus with uncharged neutrons and positively charged protons confined in a very tight space held together by the nuclear force.[1] The negatively charged electrons are confined in regions around the positively charged nucleus - sometimes the visualization of a planetary system is helpful, where the nucleus is the center and electrons are visualized to be orbiting around the nucleus in orbitals. This confinement of subatomic particles in fact leads to the quantization of physical quantities, like energy, momentum, etc. Among other details, this quantization depends on the physical size of the confining potential - just like the wavelength of nodes on a guitar string depends on the length of the string. To capture the effect of quantization, we need to study the Bohr's model of Quantum Mechanics. When this theory was developed and finally extended to Schrödinger's equation of matter waves, as we discuss in this chapter, there were some intriguing and counter-intuitive

---

[1]Neutrons and protons consist of quarks, which are fundamental particles. Interested readers are encouraged to consult the standard model of elementary particle physics.

© Springer Nature Switzerland AG 2019
H. Raza, *Freshman Lectures on Nanotechnology*, Undergraduate Lecture Notes in Physics,
https://doi.org/10.1007/978-3-030-11733-7_3

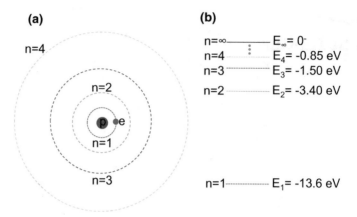

Fig. 3.1 Bohr's model of hydrogen atom. **a** The orbital picture shows the real space localization of electrons around the nucleus. **b** The energy domain picture conveys information about the energy levels of various orbitals and is quite useful for analyzing the energy exchange e.g. during photon emission and absorption

features. These may be understood by using a probabilistic approach towards this new kind of mechanics. In Quantum Mechanics, one describes the *probability* of an event, similar to the probability of finding a head or a tail in coin tossing or probability of finding a number between one to six in rolling a dice.

As a side note, the idea of atoms being the building blocks of matter is well accepted now and we take it for granted. However, scientific community in the early twentieth century did not accept this picture completely. In fact, one of the proponents of atomic theory, Ludwig Boltzmann, committed suicide due to severe depression, generally thought to be due to the non-acceptance of his scientific work related to the atomic theory of matter.

3.1 Atomic Model

Consider the planetary model of atoms, where electrons orbit around the nucleus consisting of neutrons and protons. Such a model for hydrogen atom is shown in Fig. 3.1a, which consists of one electron and one proton. The positive and negative charges lead to electrostatic attraction and cause the electron to move around the nucleus in circular orbitals. This is analogous to the motion of moon or a satellite around the earth in a gravitational field, for instance.

The centripetal force (2.7) and the electrostatic force (2.14) are equated as follows,

$$\frac{m_e v^2}{r} = K \frac{q^2}{r^2} \tag{3.1}$$

Simplifying for the speed of electron, one obtains,

$$v = \sqrt{\frac{Kq^2}{rm_e}} \tag{3.2}$$

Similarly, the total energy of electron is the sum of its kinetic energy and the potential energy given as (2.13) and below,

$$E = \frac{1}{2}m_e v^2 - \frac{Ke^2}{r} \tag{3.3}$$

Rearranging (3.1), one obtains,

$$\frac{1}{2}m_e v^2 = \frac{1}{2}\frac{Ke^2}{r} \tag{3.4}$$

Combining above equation with (3.3) of the total energy, one gets,

$$E = -\frac{1}{2}\frac{Ke^2}{r} \tag{3.5}$$

As particles move or waves travel, they ought to loose energy due to dissipation according to the classical/continuum/Newtonian mechanics. The natural consequence of such energy loss, if it was happening in these particles, would be the collapse of electron in the nucleus due to positive electrostatic attraction. But, we know, based on the experimental evidence, that atoms do not spontaneously collapse just like this – otherwise the universe's existence would be in jeopardy. However, one must understand that this picture is derived from the laws of Classical Mechanics. In the early part of the twentieth century, there was compelling experimental evidence that the laws of Classical Mechanics do not describe the mechanics of atoms or matter waves. That is where, Bohr introduced the idea of quantization of the angular momenta of electrons around the nucleus. He hypothesized that the angular momentum, given by (2.6), is discretized as integer multiples of the Planck's constant as follows,

$$m_e v_n r_n = n\hbar = n\frac{h}{2\pi} \tag{3.6}$$

where h is the Planck's constant, and the principle quantum number ($n = 1, 2, 3, \ldots$) represents various atomic orbitals with $n = 1$ as the lowest orbital with least energy and momentum. This equation revolutionized the thinking about the atomic scale matter. Philosophically, it states that the electrons may only have discrete angular momenta, which lead to discrete speeds and discrete radii of orbitals. Substituting the velocity from (3.4), one may obtain,

$$m_e \sqrt{\frac{Ke^2}{r_n m_e}} r_n = n\hbar \tag{3.7}$$

Solving for the radii of orbitals,

$$r_n = \frac{n^2 \hbar^2}{Ke^2 m_e} \tag{3.8}$$

Hence, the radii may have only discrete values depending on the value of n, since all the other factors are either universal constants or material parameters.[2] For $n = 1$, the radius of the first orbital is given as,

$$r_1 = 0.529 \times 10^{-10} \, \text{m} = 0.529 \, \text{Å} \tag{3.9}$$

where $1 \, \text{Å} = 10^{-10} \, \text{m}$. Equation 3.8 may be rewritten by using (3.9) as follows,

$$r_n = 0.529 \, n^2 \, \text{Å} \tag{3.10}$$

The radius increases quadratically with the principle quantum number (n). This is physically understandable, since the electrons close to the nucleus shield the positive charge experienced by the electrons away from the nucleus and hence the radius is incrementally larger due to the reduced effective potential of the nucleus.

Similarly, by using the quantized value of the radii, (3.5) for the total energy becomes,

$$E_n = -\frac{1}{2} \frac{Ke^2}{r_n} = -\frac{Ke^4 m_e}{2\hbar^2} \frac{1}{n^2} \tag{3.11}$$

For $n = 1$, the total energy of electron is,

$$E_1 = -2.176 \times 10^{-18} \, \text{J} \tag{3.12}$$

It is convenient to define a unit of energy called *electron-Volt*,[3] abbreviated as eV and given as $1 \, \text{eV} = 1.6 \times 10^{-19} \, \text{J}$. The energy in eV of an electron residing in $n = 1$ orbital becomes,

$$E_1 = -13.6 \, \text{eV} \tag{3.13}$$

Combining (3.11) and (3.13), one gets,

$$E_n = \frac{-13.6}{n^2} \text{eV} \tag{3.14}$$

---

[2]Which usually do not change in a certain environment.

[3]One electron-Volt (eV) is defined as the energy obtained by a single electron, when placed between two parallel plates with a voltage difference of one *Volt* between them.

The energy increases quadratically (with a negative sign) with increasing n. For $n = \infty$, the energy becomes

$$E_\infty = 0^-$$ (3.15)

for which, an electron is so far away from the nucleus (i.e. $r_\infty = \infty$) that its total binding energy is zero and it becomes a free electron. This energy level also leads to the concept of a vacuum level. Abstractly speaking, if an electron is taken to a vacuum level from a bound level, it becomes free.

At this time, we should also introduce the concept of occupied orbitals and unoccupied orbitals. In the case of hydrogen, there is only one electron, which is housed in the $n = 1$ orbital. In this case, the orbital may be termed as an occupied orbital. Since the rest of the orbitals do not have any electrons, they are called unoccupied orbitals. Furthermore, the highest orbital that is filled with electrons is called the highest occupied molecular orbital (HOMO) and the lowest orbital that is unoccupied is correspondingly called the lowest unoccupied molecular orbital (LUMO). In case of hydrogen atom, $n = 1$ orbital is HOMO and $n = 2$ orbital is LUMO.

When an electron transitions from one level, say $n = 2$, to another level, say $n = 1$, the energy exchange during this process may be emitted as a photon[4] with energy $\hbar\omega$ equal to the difference between the two levels as follows,

$$\hbar\omega = \Delta E = -13.6 \left[\frac{1}{n_i^2} - \frac{1}{n_j^2} \right] \text{eV}$$ (3.16)

where i and j represent two orbitals with different energy levels. Thus, if an electron transitions from an orbital of higher energy level to that of a lower energy level, the difference in energy may give rise to a photon emission. The reverse situation may correspond to a photon absorption.

The orbital picture in Fig. 3.1a is a good representation of the *real space distribution* of the electron orbitals around the nucleus, but it is usually more informative to draw the *energy domain picture* or the *energy levels* corresponding to the orbitals as shown in Fig. 3.1b. Such energy domain picture describes the physical quantities as a function of energy and this way of thinking is extremely useful to study various physical effects. For example the electron transition between various levels is quite evident from their energy values in the energy domain picture. Such processes between various quantized orbitals is the primary reason of emission and absorption at characteristic frequencies.

Human eye is sensitive to the transitions in the optical range (0.4–0.8 μm). These optical transitions with quantized frequencies or wavelengths are precisely the reason why different atoms and molecules (the building blocks of matter) have different colors! In fact, there is a group of metals called transition metals, like gold, chromium, cobalt, iron, etc, in the periodic table with distinct colors due to unique optical

---

[4]This energy exchange may take other forms as well, e.g. phonons (or heat), etc.

transitions. Every day examples of such observations are immense, e.g. the red color of roses and the yellow color of tulips owe their beauty to the quantized optical transitions of the dye molecules inside the respective pellets.

3.2   Orbitals and Quantum Numbers (n, l, m, s)

In the previous section, we have seen that the principle quantum number (n) gives rise to the quantization of energy. There are in fact more than one type of quantum numbers. The atomic nature of matter has symmetry that is tied down to these numbers.

The next in line to n is the quantum number associated with the *angular momentum* and is labeled as l. $l = 0$ corresponds to an s-orbital. $l = 1$ is a p-orbital, whereas $l = 2$ is a d-orbital and finally $l = 3$ is an f-orbital. For a given n, the angular momentum quantum number (l) varies from 0 to $n - 1$.

For each l, there is a set of quantum numbers m that correspond to the magnetic nature of the angular momentum due to the orbital motion, since a moving charged particle gives rise to a magnetic field. For every l, there are $2l + 1$ *magnetic quantum numbers* (m), given as $-(l - 1), -(l - 2), \ldots, -1, 0, 1, (l - 2), (l - 1)$.

Finally, for each combination of (n, l, m), there is a *spin quantum number* (s), which may have a value of either $+1/2$ or $-1/2$, usually called *up*-spin ($\uparrow$-spin) or *down*-spin ($\downarrow$-spin), respectively. The spin of an electron is also responsible for the magnetic properties of matter along with the quantum number m, as we will see more clearly in the next section. The spin property of electron enables occupancy of a single energy level with two electrons, one with $\uparrow$-spin and the other with $\downarrow$-spin, known as *Pauli's Exclusion Principle*. This states that no two electrons with the same spin may occupy the same orbital. In other words, no two electrons in an atom may have the same set of quantum numbers (n, l, m, s).

Consider the case of $n = 1$, for which $l = 0$ (s-orbital) and $m = 0$. Thus there may be two states with $\uparrow$-spin and $\downarrow$-spin, in which one may place two electrons for the same energy level in the absence of a magnetic field. Such states are usually represented as $(1, 0, 0, \uparrow)$ and $(1, 0, 0, \downarrow)$; or in short as $1s_\uparrow$ and $1s_\downarrow$, respectively. Thus, in a $1s$ orbitals, one may have two electrons including the spin degree of freedom.

For $n = 2$, l has two values, i.e. $l = 0$ (s-orbital), and $l = 1$ (p-orbital). For each value of l, one has a set of m values. For $l = 0$, $m = 0$ with the two spin states, resulting in the quantum states of $(2, 0, 0, \uparrow)$ and $(2, 0, 0, \downarrow)$. Whereas for $l = 1$, m has three values, i.e. $-1, 0, 1$. For each values of m and $\uparrow$-spin and $\downarrow$-spin, one obtains six states, $(2, 1, -1, \uparrow)$, $(2, 1, -1, \downarrow)$, $(2, 1, 0, \uparrow)$, $(2, 1, 0, \downarrow)$, $(2, 1, 1, \uparrow)$, $(2, 1, 1, \downarrow)$. Hence, one may have two electrons in the $2s$ orbitals and six electrons in the $2p$ orbitals, including the spin degrees of freedom. In total, there may be up to eight electrons for $n = 2$, as expected.

Similarly for $n = 3$, $l = 0$ (s-orbital), $l = 1$ (p-orbital) and $l = 2$ (d-orbital) have one, three and five values of m, respectively. Including the spin degrees of freedom, one has two $3s$-orbitals, six $3p$-orbitals and ten $4d$-orbitals, giving rise to eighteen orbitals for $n = 3$. To summarize, an s-orbital, a p-orbital, a d-orbital, and an f-orbital may have up to two, six, ten, and fourteen electrons, respectively, including the spin degree of freedom.

3.3   Probability Picture

The electron density distribution in various orbitals with distinct quantum numbers (n, l, m) are shown in Fig. 3.2. One may note that the shape of these distributions are quite different from the ones in Bohr's model. These distributions for various orbitals are different from each other too, giving rise to unique symmetry properties of the materials depending on the contributions from various orbitals. In a nutshell, these orbitals interact with each other at the atomic scale and form the basis of material properties at all length scales from nano to macro!

From this simple consideration, it is evident that one has to go beyond the Bohr's model of atoms to incorporate the wave nature of matter. Or in other words, one needs the mathematical description of the wave equation of matter, which was provided by Schrödinger. We discuss it further in the next section. Schrödinger equation not only gives quantized energy values, but also the wavefunctions, similar to the wavepackets we discussed in the previous chapter. The wavefunction is usually written as:

$$\Phi(n, l, m) = R_n(r) Y_l^m(\theta, \phi) \tag{3.17}$$

where (r, θ, ϕ) are the spherical coordinates. R is the radial part, which depends on the principle quantum number n and Y is the angular contribution, which depends on the quantum numbers l and m. Furthermore, the spin quantum number (s) has two values for the up-spin ($\uparrow$-spin) or $down$-spin ($\downarrow$-spin) states.[5]

Although the wavefunction (Φ) itself does not have any physical meaning, but its magnitude squared, given as,

$$|\Phi(n, l, m)|^2 \tag{3.18}$$

Fig. 3.2 Atomic orbitals for various quantum numbers. Density plots for few combinations of (n, l, m) are reported

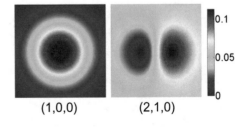

(1,0,0)   (2,1,0)

---

[5]Interested readers are encouraged to learn more about spinors.

gives the probability per unit volume in 3D[6] of finding an electron or in other words, the density of electrons, which has been plotted in Fig. 3.2 for various (n, l, m) orbitals.

Since the total probability is unity or the total density for one electron should add up to one, the volume integral over $|\Phi(n, l, m)|^2$ should be unity, given as:

$$\int |\Phi(n, l, m)|^2 \, dv = 1 \tag{3.19}$$

3.4   Wave Equation of Matter

The time independent Schrödinger equation is given as:

$$\tilde{H}\Phi = E\Phi \tag{3.20}$$

where $\tilde{H}$ is an operator called Hamiltonian, which is the sum of the kinetic energy and the potential energy. In one-dimension, $\tilde{H}$ is given as,

$$\tilde{H} = -\frac{\hbar^2}{2m}\frac{d^2}{dx^2} + U \tag{3.21}$$

which makes the Schrödinger equation as:

$$E\Phi = -\frac{\hbar^2}{2m}\frac{d^2\Phi}{dx^2} + U\Phi \tag{3.22}$$

Assuming a traveling wave[7] solution of the wavefunction as:

$$\Phi = A \, e^{ikx} \tag{3.23}$$

where A is the normalization coefficient[8] to make sure that (3.18) is satisfied. Taking the first derivative, one obtains,

$$\frac{d\Phi}{dx} = ike^{ikx} A = ik\Phi \tag{3.24}$$

---

[6]In 2D, it gives the probability per unit area, and in 1D, it gives the probability per unit length.

[7]The sum of two traveling waves in opposite direction gives a standing wave pattern.

[8]The value of the normalization coefficient should be adjusted to make the integral over the probability density equal to unity.

and

$$\frac{d^2\Phi}{dx^2} = -k^2 e^{ikx} A = -k^2 \Phi \tag{3.25}$$

since $i^2 = -1$ from the basic definition that $i = \sqrt{-1}$.
   Combining (3.22), (3.23) and (3.25), one obtains,

$$E e^{ikx} A = \frac{\hbar^2 k^2}{2m} e^{ikx} A + U e^{ikx} A \tag{3.26}$$

which simplifies to,

$$E = \frac{\hbar^2 k^2}{2m} + U \tag{3.27}$$

Substituting the (2.27) of momentum $p = \hbar k$, one obtains,

$$E = \frac{p^2}{2m} + U \tag{3.28}$$

   From (2.8), the first term of the above equation is clearly the kinetic energy and the second part is the potential energy, which in the electrostatic field is given by (2.11).
   This heuristic introduction to Schrödinger equation, which is also known as the wave equation of matter, introduces us to the probabilistic nature of the matter waves in the form of wavefunctions. The take home message is that the solution set of this equation gives the energy values and the wavefunctions!

Problems

3.1 Calculate the energy value of $n = 2$ orbital in a hydrogen atom according to Bohr's model.
3.2 Calculate the energy value of $n = 3$ orbital in a hydrogen atom according to Bohr's model.
3.3 Calculate the energy value of $n = 4$ orbital in a hydrogen atom according to Bohr's model.
3.4 Calculate the energy value of $n = \infty$ orbital in hydrogen atom according to Bohr's model. What is the physical meaning of this value?
3.5 Calculate the Bohr radius for $n = 2$ orbital.
3.6 Calculate the Bohr radius for $n = 3$ orbital.
3.7 Calculate the Bohr radius for $n = 4$ orbital.
3.8 Calculate the Bohr radius for $n = \infty$ orbital. What is the physical meaning of this value?

3.9 If an electron makes a transition from $n = 2$ orbital to $n = 1$ orbital, what is the value of the energy exchange?

3.10 $(n, l, m) = (2, 0, 0)$ corresponds to which orbital?

Research Assignment

R3.1 Explore the automobile applications of nanotechnology. Such applications include electronics, sensors (airbag sensor, oil sensor, tire pressure sensor), fabrics, glass, paint, catalytic converter, etc. Pick a topic of your choice about how nanotechnology is affecting automobile industry, and write a one-page summary.

Chapter 4
Atomic Structure and Interactions

If one wonders about what is common amongst all things, perhaps the best answer would be that they all are made up of atoms. It is noteworthy realization that different arrangement of atoms lead to such different set of properties, behavior and character that no two living beings are alike! This thinking may further be applied to man-made and naturally occurring materials and structures. Fundamentally, by arranging atoms in various forms and configurations, one may create marvels of technology. Therefore, it is indispensable to understand how atoms interact with each other and how this arrangement affects the nanoscale properties.

4.1 Bonding

Electrons inside atoms may be divided into two categories, core electrons which are close to the nucleus, and valence electrons, which are the outermost electrons. When atoms approach each other, the valence electrons interact with each other and form chemical bonds. This truly is a femto second (10^{-15} s) process. The number of bonds per atom depends on the number of valence electrons, e.g. hydrogen has one valence electron, so it forms a single bond per atom; carbon and silicon have four valence electrons, therefore they form four bonds per atom.

The governing principle behind atomic interaction is to lower the total energy of the system. In other words, atoms form chemical bonds with the other atoms to lower the total energy. The chemical bonds may be of multiple types, e.g. metallic, covalent, ionic, etc. In this chapter, we do not go into the detail about the character of each type of bond, rather focus on the overall idea.

Consider two hydrogen atoms reacting to form a hydrogen molecule, as follows:

$$H + H \rightarrow H_2 \tag{4.1}$$

© Springer Nature Switzerland AG 2019

H. Raza, *Freshman Lectures on Nanotechnology*, Undergraduate Lecture Notes in Physics,
https://doi.org/10.1007/978-3-030-11733-7_4

Fig. 4.1 Chemical bonding in a hydrogen molecule. Two hydrogen atoms interact to form a hydrogen molecule. The chemical reaction is shown by using **a** the orbital picture, and **b** the energy domain picture

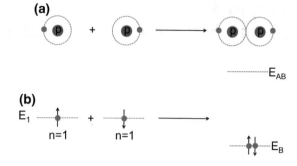

This chemical process may be described in the orbital picture for better visualization, schematically shown in Fig. 4.1a. The atomic orbitals of the two hydrogen atoms interact with each other to form a single molecular orbital over the two nuclei, which results in a chemical bond formation between the two atoms. The single molecular orbital indeed holds the two electrons, with opposite spins. The orbital picture is very helpful to see how atoms are arranged and how electron densities are distributed around the nuclei. Although very illustrative, the orbital picture does not convey the complete information about the feasibility and other important properties of the chemical reaction. In fact, one often uses even a shortened version of the orbital model by displaying atoms as spherical balls and bonds as sticks, known as *ball and stick model*.

A more useful description of such a process is the representation in the energy domain, which leads to a better understanding of the energetics as shown in Fig. 4.1b. When two atomic orbitals interact with each other, they actually form two molecular orbitals, one bonding which is lower in energy than the atomic orbitals and the second anti-bonding which is higher in energy than the atomic orbitals. These two atomic orbitals may hold two electrons of opposite spin each, thereby totaling four electrons. After the reaction, the resulting species should be able to hold the same number of electrons as the atomic orbitals. That is why the two atomic orbitals always give two molecular orbitals - albeit one lower in energy and the other one higher.

In energy level description, such atomic and molecular levels are shown in Fig. 4.1b, where bonding and anti-bonding energy levels are labeled as E_B and E_{AB} respectively. In the present example, two hydrogenic atomic levels of $n = 1$ with one electron per level react to form a bonding level that has lower energy than $E_1 = -13.6\,\text{eV}$. Since electrons fill the lower energy orbitals first to decrease the total energy, both the electrons reside in the bonding level within the hydrogen molecule, nonetheless with opposite spins. The net energy reduction in this case is $2(E_1 - E_B)$ for the two electrons. This overall energy reduction is the primary reason why atoms form bonds with each other to form molecules and other complex structures.

4.2 Dimensionality

We perceive to live in a three-dimensional (3D) space. Therefore, most materials
and structures we observe are 3D. However, there are plenty of lower-dimensional
materials, as shown in Fig. 4.2 by using ball and stick models, which are further
discussed in this section.

Strictly speaking, a structure is 3D if its physical dimensions span from $-\infty$ to
$+\infty$ in all three dimensions. But for all practical reasons, if the size is greater than tens
of de Broglie wavelength in all three dimensions i.e. there are no significant matter
wave effects, the material may be assumed to be 3D. For example, in Aluminum,
the de Broglie wavelength is about 5 nm. Therefore, a cube of Aluminum with side
50 nm or more may easily be categorized as a 3D material. Another example is that of
graphite, commonly found in lead pencils. The cylindrical piece of graphite in your
pencil may be easily classified as a 3D material. If one looks at the atomic scale,
graphite consists of two-dimensional (2D) monolayers of carbon atoms known as
graphene, which are stacked on top of each other.

In 2D materials, the size in one dimension is comparable to or smaller than the
de Broglie wavelength. One may imagine extracting graphene from graphite. Experimentally, a graphene sample may easily be tens of microns wide in the lateral

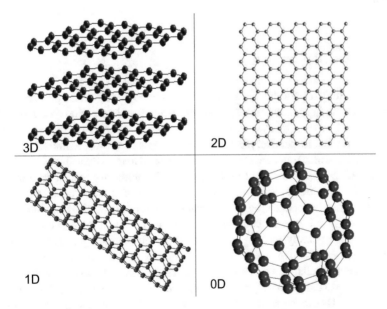

Fig. 4.2 Dimensionality illustrated by using ball and stick models. Three-dimensional structure
have macroscopic features, e.g. a graphite sample or a silicon wafer. Two-dimensional nanomaterials include graphene and atomically thin films. One-dimensional (1D) nanostructures include
carbon nanotubes (CNT) and silicon nanowires. Atoms may be arranged in zero-dimensional (0D)
nanostructures, e.g. a C_{60} molecule or Bucky ball

directions and of course graphene is single atom thick in the third direction - thus making it an ideal 2D material.

One may further imagine taking a narrow, say 3 nm wide, ribbon of carbon atoms out of graphene and roll it in the width direction to form a tube with about 1 nm diameter, widely known as a carbon nanotube (CNT). Such a CNT is an example of a one-dimensional (1D) material, for which the size in two dimensions is comparable or smaller than tens of de Broglie wavelength. In lab, one may routinely grow $1-2$ nm diameter CNTs, which are few tens of μm long.

Finally, a zero-dimensional (0D) material is the one that has size comparable to or less than tens of de Broglie wavelength in all three dimensions. Examples of such structure include C_{60} or Bucky ball and in fact all organic molecules, nanocrystals, quantum dots, etc. Such structures are synthesized routinely in the lab and in many cases are found in the nature.

Although one may follow a rigorous classification based on the comparisons with the de Broglie wavelength, usually, if one ponders about a structure, it is straightforward to decide the dimensionality. Just by observing the number of dimensions in which the structure or the material has a large size, one may assign the dimensionality. So the next question is how the atomic arrangement affects physical and chemical properties of a material? Although, we try to address this question in the next chapter, a brief description of the concept of unit cell for classification of various materials is addressed next.

4.3   Unit Cells

In order to study a periodic sinusoidal wave, one may describe its properties very elegantly by stating its amplitude, wavelength and the time period as discussed in Sect. 2.2. This concept of wavelength and time period summarizes the repeatability in space and time, respectively. Similarly, in crystals of various materials, one may define a region called unit cell, which may be repeated to reproduce the whole crystal. Consider 1D crystal shown in Fig. 4.3a, where the atoms are placed on a uniform lattice with the lattice spacing a. In this case, a unit cell may consist of a single atom which may be repeated by a unit distance a to reproduce the 1D atomic crystal. However this is not the only combination of the unit cell and the unit distance. One may have done equally well by defining a unit cell consisting of two atoms and placing them $2a$ distance apart. Thus for a periodic atomic arrangement, one may have infinite unit cells. The one with the smallest physical dimensions (length in 1D, area in 2D and volume in 3D) is called the *primitive unit cell*. Consider a real-life example of a 1D graphene nanoribbon in Fig. 4.3b. The unit cell is more detailed, consisting of 18 carbon atoms, which repeats itself on the 1D lattice to generate the nanoribbon crystal.

Similarly, for the graphene structure in Fig. 4.3c, the unit cell consists of two carbon atoms, which may be repeatedly placed over a 2D plane with appropriate spacing to reproduce the honeycomb crystal. For the 3D silicon and NaCl structures

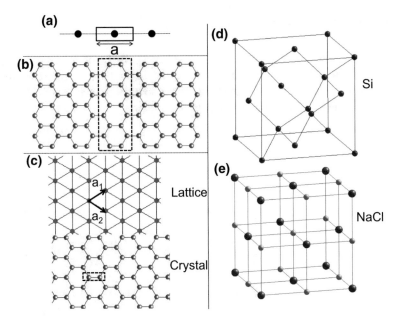

Fig. 4.3 Concept of a unit cell in various dimensions. **a** 1D crystal, and unit cell. **b** Unit cell of an armchair GNR. **c** 2D graphene lattice, unit cell, and crystal. **d** Silicon crystal in 3D. **e** NaCl crystal in 3D

shown in Fig. 4.3d, e, respectively, the unit cell is repeated over a 3D lattice, hence becomes a little complicated, but the basic concept stays the same as in 2D or 1D.

In short, for periodic crystals, the concept of a unit cell helps to compact the information about the physical arrangement of atoms into a few atoms of the unit cell and the lattice.

Problems

4.1 If two identical atoms have an energy level at $-10\,\mathrm{eV}$, and after bonding, the molecule has energy levels at -5 and $-15\,\mathrm{eV}$, which of the two levels belong to the bonding and the anti-bonding orbitals? Explain your answer.

4.2 Look up the de Broglie wavelength of electrons in a metal of your choice.

4.3 Look up the de Broglie wavelength of electrons in a semiconductor of your choice.

4.4 Look up the de Broglie wavelength of electrons in an insulator of your choice.

4.5 Pick a 1D nanomaterial and identify its primitive unit cell.

4.6 Pick a 2D nanomaterial and identify its primitive unit cell.

4.7 Pick a 3D material and identify its primitive unit cell.

Research Assignment

R4.1 Nanotechnology is impacting aerospace technology immensely. Consider lightweight materials, jet engines, composites, radiation sensors, radiation hardened electronics, etc. Pick a topic of your choice about how nanotechnology is impacting aerospace technology, and write a one-page summary.

Chapter 5
Electronic Structure

Given that all matter consists of atoms, there are only 90 naturally occurring elements on the planet Earth.[1] However, if one looks at the naturally occurring or even man-made materials in daily life, there is an enormous range of structures and properties. Such a diversity is precisely due to the various arrangements at the atomic scale of an otherwise finite set of elements. In the previous chapter about the atomic arrangement, we motivated the reader by stating that *it is indispensable to understand how atoms interact with each other and how this arrangement affects the nanoscale properties.* In this chapter, we focus on the role that atomic arrangement in various dimensions plays in determining the physical and chemical properties of a material.

5.1 Electronic Spectrum

By looking at the rainbow on a fine day, one may spot various colors arranged in an elegant manner. In more precise terms, what one is looking at is the wavelength or frequency distribution of an otherwise ordinary sunlight. Since wavelength (λ), frequency (f) and energy (E) are inter-related,[2] the constituent colors in a rainbow or the spectrum of the sun light may be studied by analyzing its various energy components.

For the electronic structure of any material, the situation is no different! Primarily, one is interested in the electronic spectrum in the energy-domain picture, i.e. how various quantities are distributed as a function of energy. To illustrate further, consider the example of the distribution of energy levels in an atom. As discussed in Chap. 3, according to Bohr's model, energy levels of an atom are quantized, and the corresponding energy values are given as:

---

[1] Elements with atomic numbers 1 through 92 (except elements with atomic numbers 43 and 61) occur naturally.

[2] $E = hf = hc/\lambda$, where h is Planck's constant and c is the speed of light.

© Springer Nature Switzerland AG 2019
H. Raza, *Freshman Lectures on Nanotechnology*, Undergraduate Lecture Notes in Physics,
https://doi.org/10.1007/978-3-030-11733-7_5

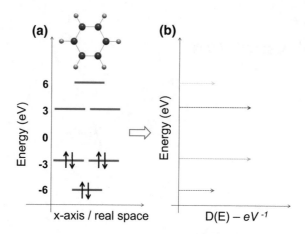

Fig. 5.1 Concept of the density of states (*DOS*). **a** $E(x)$ diagrams for the energy levels may be replaced by the *DOS* plots, which shows the amplitude of the electronic spectra as a function of energy. **b** For a benzene molecule with HOMO (highest occupied molecular orbital) and LUMO (lowest unoccupied molecular orbital) energy levels. The *DOS* plot consists of discrete energy levels corresponding to the molecular orbitals

$$E_n = -\frac{13.6}{n^2} \text{ eV}, \quad \text{where } n = 1, 2, 3, \ldots \quad (5.1)$$

These are the possible locations (in energy) where electrons may reside. By using proper book-keeping, one may fill the energy levels according to Hund's rule[3] and Pauli's exclusion principle. It may seem convenient for one atom, but it is certainly not tractable for more than a few atoms! One has to devise a more convenient way of analyzing the energy states, which tells us about the number of such energy levels distributed over energy. Such a quantity is called Density of States (*DOS*) and is defined as follows,

$$D(E) = \sum_{n=1}^{\infty} \delta(E - E_n) \quad (5.2)$$

where $\delta(E - E_n)$ is called the Dirac's delta function[4] and is non-zero only for $E = E_n$. Furthermore, the integral over energy (i.e. the area under the curve) for this function is unity,

$$\int \delta(E - E_n) \, dE = 1 \quad (5.3)$$

which physically states that each energy level may hold one electron per spin. Thus, for each n corresponding to an energy level, the density of states for hydrogen atom has a delta function. From the above equation, it is evident that the units of $D(E)$

---

[3]Interested reader is encouraged to learn more about this rule.
[4]Kronecker's delta function is a different one.

should be per unit energy, i.e. /eV. Concisely, *DOS* is the number of available states per eV of energy.

5.2 *DOS* and Dimensionality

In the previous chapter, we emphasized that the dimensionality of a material changes its physical and chemical properties. In this section, we study the effect of dimensionality on the *DOS* of materials, more broadly referred to as the electronic structure of materials. Indeed, the term electronic spectrum used in the previous chapter is synonymous with the electronic structure.

5.2.1 0D Materials

In Sect. 5.1, we discussed that the arrangement of allowed energy states for a hydrogen atom may be described by the density of states consisting of delta functions at the location of energy levels. In fact, this feature is characteristics of 0D materials, e.g. organic molecules, nanocrystals, quantum dots, etc. Consider the case of 0D benzene molecule. Just like hydrogen atom has atomic orbitals with discrete energy levels, benzene molecule has molecular orbitals and corresponding energy levels, which are schematically shown in Fig. 5.1a. The density of states reflecting the location of energy levels is also shown in Fig. 5.1b, which consists of delta functions given by (5.2).

In analogy with the real space picture of filling the orbitals with electrons according to the Pauli's exclusion principle and the Hund's rule, one may form an equivalent picture of filling the density of states. We discuss this further in Sect. 5.4.

5.2.2 1D Materials

1D nanomaterials may be thought of as a pile of 0D materials arranged in one-dimension. The corresponding energy levels are schematically shown in Fig. 5.2a. In Sect. 5.1, we noticed that the interaction of discrete orbitals leads to new energy states. For the 1D electronic spectrum, it turns out that the delta function feature of the 0D density of states is retained, which in this case are called van Hove singularities. Additionally, new states emerge off the delta function. The *DOS* for a 1D structure like carbon nanotube is given as,

$$D(E) \sim \sum_{n=1}^{\infty} \frac{1}{\sqrt{(E - E_n)}} \qquad (5.4)$$

Fig. 5.2 Electronic structure of 1D structures. **a** Energy levels for a 1D structure as a function of real space are schematically shown. **b** 1D retains some of the features of 0D like spikes in the *DOS*, but also has a tail-like feature due to extra available states

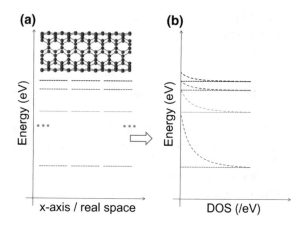

and is schematically shown in Fig. 5.2b.

5.2.3   2D and 3D Materials

A two-dimensional silicon oxide film and a three-dimensional silicon crystal are shown in Fig. 5.3a, b, respectively. One may extend the analysis that 2D *DOS* is an aggregate of multiple interacting 1D structures. In this analogy, the 1D *DOS* shown in Fig. 5.2b leads to the step function *DOS* shown in Fig. 5.3c in the sense that there are more states as compared to the 1D case. One may write the 2D *DOS* as a sum of multiple unit step functions as follows,

$$D(E) = \sum_{n=1}^{\infty} u(E - E_n) \tag{5.5}$$

where a unit step function $u(E - E_n)$ is defined to be zero below E_n and unity above E_n.

Extending the discussion to a 3D material, the *DOS* is given as follows:

$$D(E) \sim \sum_{n=1}^{\infty} \sqrt{(E - E_n)} \tag{5.6}$$

which is schematically shown in Fig. 5.3c as a dashed line. 3D *DOS* is a monotonically increasing function of energy above a certain energy E_n.

We have given simple examples of typical *DOS* plots in 0D, 1D, 2D, and 3D. In general, the *DOS* plots may be more detailed and complicated than the simple examples shown here. Consider graphene for example, which is a 2D material and has a linear *DOS* instead of the unit step function for an otherwise typical 2D system.

Fig. 5.3 *DOS* plots for 2D and 3D. **a** 2D silicon oxide film, and **b** 3D silicon crystal are shown. **c** For a 2D structure like SiO_2 film, the tail features of 1D *DOS* have been *filled* by additional states, giving rise to a unit step function. Thus the *DOS* plot becomes staircase. Further increasing the number of states in 3D leads to monotonically increasing trend in the *DOS*, e.g. that of silicon

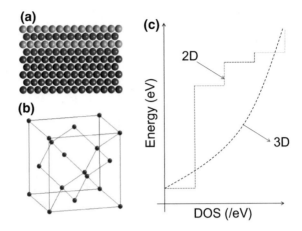

5.3 Band Picture

As discussed in the previous section, starting with the 0D *DOS*, one may form a continuum of states in 1D, 2D or 3D materials. Such a group of states is called a band of states or simply a band. The concept of density of states is very useful in understanding the energy level distribution i.e. counting the energy states to form a distribution $D(E)$. However, it is incomplete in the following sense.

In the second chapter, we learned that in order to describe any phenomenon, one has to describe it as a function of time (t) and real space (x), i.e. (t, x) coordinates. Later, the concept of transformation was highlighted, where one may transform the time domain information into the frequency domain or equivalently in the energy domain (E). However, the real space information (x) can also be transformed into momentum space (p) or equivalently wavevector[5] space (k). Thus a phenomenon may be represented in either of the following equivalent forms without any loss of information,

(t, x)

(t, k)

(E, x)

(E, k)

It is this last set of coordinates, i.e. (E, k) that is particularly useful, since energy and momentum or wavevector are related by a parabolic expression as,

$$E = U + \frac{p^2}{2m} = U + \frac{\hbar^2 k^2}{2m} \tag{5.7}$$

where U is the potential energy and m is the mass. Hence, the complete information about the electronic properties is contained in the $E(k)$ diagrams for a crystalline

---

[5]$p = \hbar k$.

Fig. 5.4 Band structure.
a *DOS* plot. **b** $E(k)$ diagrams
have k-resolved information
as well. An $E(k)$ plot thus
has both the time-domain
and real-space information

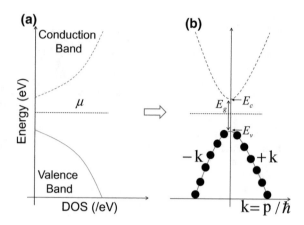

material. Once $E(k)$ plots are known, all other properties, including the density of states, may be calculated.

Consider Fig. 5.4a, b, where the *DOS* and $E(k)$ plots are respectively shown for two bands. Let us assume that the bottom band is mostly filled with electrons and the top band is mostly empty. In such an arrangement, these bands are called valence band and conduction band, respectively. In terms of (5.7), the $E(k)$ relationship for the parabolic conduction band may be represented by,

$$E = E_c + \frac{p^2}{2m_c} = E_c + \frac{\hbar^2 k^2}{2m_c} \tag{5.8}$$

where E_c is the conduction band edge and m_c is the mass of the conduction band. m_c could in general be different form the free electron due to different crystal environment. Some crystals facilitate electron transport, whereas some hinder their movement, giving rise to lighter and heavier masses, respectively. For this very reason, k is also called crystal momentum, and in fact may be quite different from that of a free electron.

The $E(k)$ relationship for the parabolic valence band may be written as,

$$E = E_v + \frac{p^2}{2m_v} = E_v + \frac{\hbar^2 k^2}{2m_v} \tag{5.9}$$

where E_v is the valence band edge and m_v is the mass of the valence band, which is in fact negative. Again, it is not to be taken as the actual mass, but rather an effective mass felt inside a crystal. For this same reason, $m_{c,v}$ are also called the effective masses of the conduction and the valence bands, respectively. The difference between E_c and E_v is called a band gap E_g Fig. 5.4b, which we discuss further in Sect. 5.5.

One may also observe in Fig. 5.4 that the majority of the states have a non-zero k value. This value is related to velocity or momentum as follows,

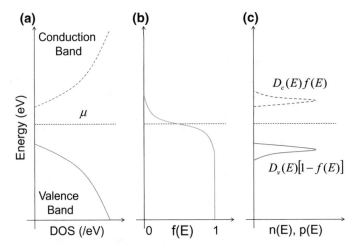

Fig. 5.5 Fermi's function. **a, b** Coupling *DOS* with the Fermi's function, one may obtain **c** electron and hole distribution functions

$$p = mv = \hbar k \tag{5.10}$$

Thus, electrons inside a crystal, irrespective of the dimensionality, move back and forth, in $\pm k$ directions all the time! However, under equilibrium condition, the number of electrons traveling one-way is balanced by the number of electrons traveling in the opposite direction. Of course, the situation may be perturbed e.g. by applying a bias, in which case the electrons start to flow in one certain direction, leading to a net current flow – a phenomenon which we discuss in the next chapter.

5.4 Fermi-Dirac Statistics

Photons, being spin-1 particles, follow Bose-Einstein statistics as discussed in the second chapter. Electrons, being spin-1/2 particles, follow another kind of statistics called Fermi-Dirac statistics and thereby have a different distribution function. Therefore, the next question is that given the *DOS* distribution shown in Fig. 5.5a, how can one find the number of electrons residing in a region? For a large number of electrons, it becomes quite tedious to keep track of electrons. This is where a new thermodynamic quantity called the Fermi's function,[6] is quite helpful. Fermi's function gives the probability of finding an electron as follows,

---

[6]The symbol for Fermi's function $f(E)$, emphasizes that it is always a function of energy. One should not confuse Fermi function's symbol with that of the frequency. Furthermore, a subscript is sometimes used in the symbol as well, e.g. $f_0(E), f_1(E)$ or $f_2(E)$, etc, to cater to specific biasing conditions.

$$f(E) = \frac{1}{1 + e^{(E-\mu)/k_B T}} \tag{5.11}$$

where k_B is the Boltzmann's constant. The range of $f(E)$ is $[0,1]$.[7] The chemical potential (μ) is defined as an energy level for which all states below are filled and above are empty at an absolute temperature of zero. $k_B T \approx 25\,\text{meV}$ at room temperature, for which $f(E)$ is shown in Fig. 5.4b.

Since $f(E)$ gives information about the probability of electron occupancy of a state and $D_c(E)$ gives the density of conduction band states, their product gives the electron distribution function per spin as a function of energy inside the conduction band as shown in Fig. 5.5c and is given as follows,

$$n(E) = D_c(E)f(E) \tag{5.12}$$

The total number of electrons, per spin, inside the conduction band is then given as,

$$N = \int_{E_c}^{\infty} n(E) = \int_{E_c}^{\infty} D_c(E)f(E) \tag{5.13}$$

For the valence band, the same procedure may be carried out for calculating the number of electrons. However, it is not a useful approach because for such a band, it is the absence of electrons that is much more meaningful. In other words, in a bowl full of water, the bubbles (or voids of water) contain the same information. In fact, such an absence of electron is called a hole, which has a positive charge of the same magnitude as that of an electron. Incidentally, due to residing in a valence band, it has a negative effective mass.

The probability of finding a hole or the absence of an electron per spin is given as,

$$1 - f(E) = 1 - \frac{1}{1 + e^{(E-\mu)/k_B T}} \tag{5.14}$$

and the hole distribution function per spin inside the valence band, shown in Fig. 5.5c, is given as follows,

$$p(E) = D_v(E)[1 - f(E)] \tag{5.15}$$

whereas the total number of holes per spin are,

$$P = \int_{-\infty}^{E_v} p(E) = \int_{-\infty}^{E_v} D_v(E)[1 - f(E)] \tag{5.16}$$

One may independently control the number of electrons or holes in a given material by changing the chemical potential. Such a change may be induced chemically by incorporating *dopant* atoms, a process called doping, or by applying an external

---

[7]Physically, $f(E)$ gives the probability of finding an electron at a given energy. Therefore, the range $[0,1]$ includes the numbers 0 and 1, whereas $(0,1)$ does not include the numbers 0 and 1.

voltage. Apart from this, one should note that at room temperature, due to thermal excitations, electrons from the valence band transition to the conduction band, thus electron-hole pairs are formed. Of course, in this case, the number of electrons and holes are equal.

5.5 Material Classification

Based on the parameters derived from the band picture or $E(k)$ diagrams shown in Fig. 5.6, one may classify materials into different categories. The band gap (E_g) is defined as the difference in the conduction and valence band edges, i.e. $E_c - E_v$. If the band gap is on the order of $3-8$ eV, the material is an insulator as shown in Fig. 5.6a. In such materials, it is difficult to thermally generate electron-hole pairs due to a large energy gap between the valence band and the conduction band. Furthermore, doping does not help much in creating electrons or holes either and the material remains non-conducting. Examples of insulators include most oxides, nitrides, etc, e.g. SiO_2, Al_2O_3, Si_3N_4, HfO_2, ZrO_2, MgO etc. It is noteworthy that the band gap of SiO_2 is about 8.9 eV.

The second class of materials is called semiconductors, for which the band gap is in $0.1-3$ eV range as shown in Fig. 5.6b. The semiconductor band gap is more accessible than that of insulators for thermal energy, which is on the order of 25 meV at room temperature. In such materials, not only the thermal electron-hole pairs are

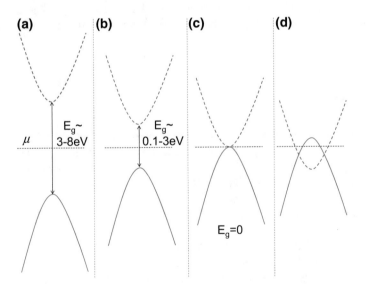

Fig. 5.6 Classification of materials. Based on the band gap, materials may be classified into **a** insulators, **b** semiconductors, **c** semimetals and **d** metals, due to their distinct electronic properties

generated, but dopant atoms are also introduced to increase the electron density or hole density independently, generating n-type or p-type semiconductors. The most well-known material in this class is perhaps Silicon, which has a band gap of 1.12 eV at room temperature and is a work-horse of today's integrated circuits. Also, silicon carbide (SiC) has a band gap of about 2.3–3 eV at room temperature and is widely used in power electronic devices. The second most important set of semiconductors consists of a combination of group III and group V (called III-V semiconductors) or group II and group VI (called II-VI semiconductors). This set of semiconductors is important for optoelectronic applications, e.g. light emitting diodes, photodetectors, etc. To be useful for such applications, a semiconductor should have a direct band gap, i.e. the top of valence band and the bottom of the conduction band should be at the same k value as shown in Fig. 5.6b. However, if this is not the case, e.g. in Si and SiC, the semiconductor is said to have an indirect band gap, which may not emit light, although still may absorb it, albeit not so efficiently. We further discuss this phenomenon in Chap. 8.

The third class of materials is called semimetals and they have a zero band gap as shown in Fig. 5.6c, for which the conduction and valence bands just meet at one point, but do not overlap! This gives rise to a large amount of thermally degenerated electron-hole pairs. Lead is an example of such a material, as well as bilayer graphene. The monolayer graphene is in fact also a semimetal but the $E(k)$ diagram is linear instead of a parabola, which gives rise to other interesting properties and applications.

Most commonly found metals, e.g. Au, Al, Ag, Cu, etc, have overlapping conduction and valence bands, where the band gap is *absent*, i.e. there is no concept of a gap, not even zero around the chemical potential[8] as shown in Fig. 5.6d. Of course, there are other kinds of phases, e.g. superconductors, Luttinger liquids, Wigner crystals, etc, which are beyond the scope of this book – however, may still have many practical applications.

Problems

5.1 For a parabolic band, plot the approximate density of states for a 1D structure.

5.2 For a parabolic band, plot the approximate density of states for a 2D structure.

5.3 For a parabolic band, plot the approximate density of states for a 3D structure.

5.4 What is the physical meaning of $f(E)$?

5.5 What is the physical meaning of $[1 - f(E)]$?

5.6 Calculate the value of the Fermi's function at $E = 0$ for $\mu = 0$, and $k_B T = 25$ meV.

5.7 Give an example of an insulator. Comment on its band gap.

5.8 Give an example of a semiconductor. Comment on its band gap.

5.9 Give an example of a semi-metal. Comment on its band gap.

---

[8]There are indeed band gaps in metals, but these are away from the chemical potential, and hence do not play an active role in determining the material properties.

5.10 Give an example of a metal. Comment on its band gap.

5.11 Comment on the velocity of electrons for a material with linear bandstructure like that of a monolayer graphene.

Research Assignment

R5.1 Nanobiotechnology and bionanotechnology are important aspects of nanotechnology. This area is expected to revolutionize nanomedicine. The medical applications include equipment sanitation, implant disinfection, lab on chip, medical implant, tissue engineering, drug screening, drug delivery, bionanosensors, nanobiosensors, pathogen detection, medical imaging (e.g. fMRI, functional medical resonance imaging), targeted drug delivery, dentistry (nanocomposites, nanoclusters, implants, bone replacement cements, nano-needles), ophthalmology, surgical robots, artificial organs, etc. Pick a topic of your choice about how nanotechnology is affecting healthcare in contemporary society, and write a one-page summary.

Chapter 6
Nanoelectronics

Nanoelectronics is the study of how electrons flow through materials and devices at the nanoscale. By controlling the electron flow in a predictable manner, one may fabricate useful devices ranging from computers, digital cameras, chemical and biological sensors, to LEDs (light emitting diodes), LASER (light amplification by stimulated emission of radiation) devices, solar cells, thermoelectric devices, etc.

In this context, Moore's law has been a driving force behind micro and nanoelectronics for the past five decades or so. This law has been successfully able to predict the doubling of the transistors in the integrated circuits every two year or so. This trend has resulted in tremendous device scaling, where the device size has been reduced from few microns to tens of nanometers. Given that the device size now is in the nanoscale regime and certainly less than the de Broglie wavelength in silicon, one has to understand the charge transport through these devices by understanding the quantum transport.

6.1 Quantum Transport

The generic device structures are shown in Fig. 6.1a, b, c. The channel is contacted by the source and the drain, thereby making it a two-terminal device as shown in Fig. 6.1a. Gate voltage may be applied through an additional gate in the three-terminal FET (field effect transistor) configuration as shown in Fig. 6.1b. There is a dielectric between the gate and the channel, which electrically isolates but electrostatically couples the gate with the channel. One may also visualize a two-terminal device with additional control or stimulus in the form of photons or molecules as shown in Fig. 6.1c. The former would be an example of a photodiode or a solar cell, whereas the latter falls in the category of a chemical or a biological sensor.

© Springer Nature Switzerland AG 2019
H. Raza, *Freshman Lectures on Nanotechnology*, Undergraduate Lecture Notes in Physics,
https://doi.org/10.1007/978-3-030-11733-7_6

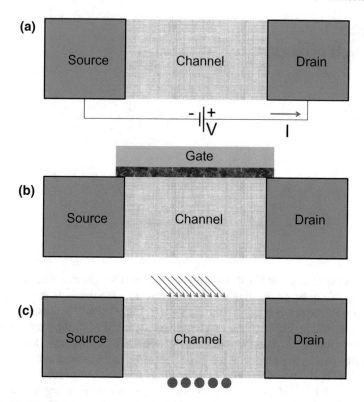

Fig. 6.1 Device structure. **a** A two-terminal device with source and drain contacts. **b** A three-terminal device with a gate contact. **c** A device may have additional stimuli in the form of external molecules depicted by the circles or incident photons depicted by the arrows

In energy domain, the channel may be represented by the density of states (*DOS*) for the conduction and the valence bands as shown in Fig. 6.2a. The group velocity is defined as,

$$v_g = \frac{d\omega}{dk} = \frac{1}{\hbar} \frac{dE}{dk} \tag{6.1}$$

For the parabolic conduction band and valence band shown in Fig. 5.4b and given by (5.8) and (5.9), respectively, the group velocity is linear with a positive or a negative value depending on the slope of the $E(k)$ band structure as shown in Fig. 6.2b. The energy resolved transmission $T(E)$ is then given as the product of the *DOS* and the group velocity v_g as shown in Fig. 6.2c. While we introduce the concept of transmission here under equilibrium conditions, what we need is the non-equilibrium transmission to understand the quantum transport. This may be calculated by using more advanced methods like non-equilibrium Green's function (NEGF) formalism. The detailed discussion of these methods is beyond the scope of this book, however here we introduce a heuristic approach to understanding the quantum transport through nanodevices.

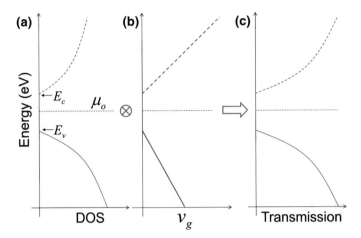

Fig. 6.2 Transmission. The product of **a** *DOS* and **b** velocity at each energy point gives **c** the energy-resolved transmission

It is important to understand how the states inside the channel are filled or emptied according to the contact's chemical potential. One may assume that the contacts stay in equilibrium even when the channel is out of equilibrium, and hence use the equilibrium statistical mechanics for the contact region in the form of Fermi's function [$f(E)$] for the spin-1/2 particles. Fermi's function gives the probability of occupancy of a state by an electron, and hence has a range of [0, 1]. The Fermi's function for the chemical potential μ_o at temperature T (in K) is given as,

$$f_o(E) = \frac{1}{1 + e^{(E-\mu_o)/k_B T}} \tag{6.2}$$

where k_B is Boltzmann's constant, and $k_B T$ is about 25 meV at $T = 300$ K. At $E = \mu_o$, the Fermi's function gives a probability of 1/2, irrespective of the temperature. By applying a bias, non-equilibrium conditions are established. Conceptually, one may imagine that the chemical potential of one contact (say drain) is changed with respect to the other contact (say source). Correspondingly, the contact chemical potential may be moved up or down in energy. Since the source contact is usually connected to ground, the source chemical potential (μ_1) equals the equilibrium chemical potential (μ_o). On the other hand, the positive bias (V_d) applied at the drain contact results in the drain chemical potential ($\mu_2 = \mu_o - qV_d$) to shift down as shown in Fig. 6.3. One useful analogy here may be of water level in a container, which moves down when the water is removed.

In the Landauer's picture of quantum transport, the channel contributes through the channel transmission $T(E)$ as shown in Fig. 6.4a, whereas the contact contributes through the contact's Fermi functions. The difference of the two contact's Fermi functions is shown in Fig. 6.4b. The product of the transmission and the difference of the Fermi's functions is also shown in Fig. 6.4c, which gives rise to the electrical

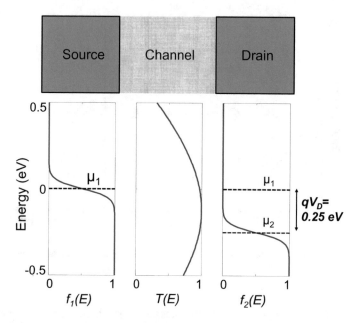

Fig. 6.3 Electrical transport. Non-equilibrium condition established by the difference in chemical potentials between the source and the drain contacts

Fig. 6.4 Landauer's picture. The product of transmission and the Fermi function difference gives rise to a net current

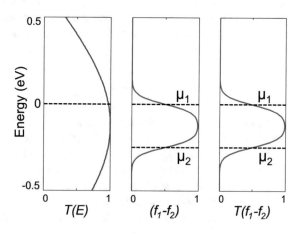

conduction. The current through the channel (including the spin degeneracy of 2) is thereby given as,

$$I = \frac{2q}{h} \int_{-\infty}^{+\infty} dE\, T(E)\left[f_1(E) - f_2(E)\right] \qquad (6.3)$$

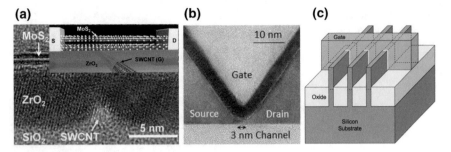

Fig. 6.5 Nanotransistor. **a** CNT transistor. **b** Silicon transistor with 3 nm channel length. **c** FinFET. 22 nm trigate FinFET structure with multiple fins [1, 2] (courtesy of Intel)

The ideal contacts are reflection-less with perfect matching at the contact-channel interface, resulting in unity transmission. The current thus becomes,

$$I = \frac{2q}{h} \int_{-\infty}^{+\infty} dE \left[f_1(E) - f_2(E) \right] \tag{6.4}$$

For small bias ($V \ll k_B T / q$), it may be shown that the above equation simplifies to,

$$I = \frac{2q^2}{h} V_d \tag{6.5}$$

giving a conductance of $2q^2/h$, which equals $77.48 \, \mu S$[1] including the spin degree of freedom per band (or state). Since the conductance comprises of universal constants only, the $2q^2/h$ is also called the *quantum of conductance*.

6.2 Some Advanced Topics

6.2.1 Nanotransistors: A Research Perspective

In this section, we discuss various examples of nanoelectronics to give the reader a feel of where the field of nanoelectronics is headed.

A nanotransistor based on Molybdenum disulfide (MoS_2) channel and a single wall carbon nanotube (SWCNT) gate is shown in Fig. 6.5a. The molybdenum disulfide channel is contacted by source and drain contacts, whereas the 1 nm gate is formed by using the CNT as the gate, which is separated from the channel by using zirconia (ZrO_2). The electron microscope viewgraphs with the side view are shown in Fig. 6.5a, whereas the schematic is reported as the inset.

---

[1]Usually approximated to $80 \, \mu S$.

Fig. 6.6 Flash memory. A storage node, known as floating gate, in the gate stack is used for storing charge

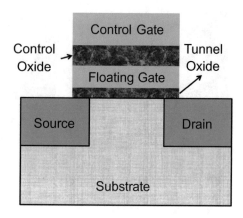

In Fig. 6.5b, the cross section viewgraph of a silicon transistor with the channel length of about 3 nm is shown. While the gate here is V shaped, it may have different shapes in other transistors, e.g. planar, round, fin (as we discuss later), etc. Since the gate mostly consists of metal, the dielectric (shown by dark color between the gate and the channel) is usually some kind of oxide (but not always), and the channel is semiconducting. This transistor is called MOSFET (metal-oxide-semiconductor FET), which has become the workhorse of modern day digital IC chips. It is instructive to note that the real structures have a lot of line and edge roughness.

At the nanoscale, it is quite difficult to have a complete electrostatic control through the gate contact over the whole channel region. A tri-gate transistor (commercially known as Fin-FET due to the shape) has been widely used for the improved gate control. One such FinFET is schematically shown in Fig. 6.5c.

6.2.2 Non-volatile Memory

It is imperative to have a defect free gate dielectric and the dielectric-channel interface in MOSFET technology. However, the gate stack may be engineered to have defects or imperfections for niche applications. One such example is the non-volatile memory device as shown in Fig. 6.6. The gate stack has an additional floating gate layer, separated from the channel and the gate by a tunnel dielectric/oxide and a control dielectric/oxide, respectively. Under the appropriate polarity and magnitude of the control gate voltage, charge may be stored in this floating gate, thereby enabling two states. The storage of charge in the storage node of the floating gate may lead to a low-resistance state and the absence of charge results in a high-resistance state, and vice versa.

Traditionally, the floating gate has been poly-silicon, whereas the substrate has been silicon. There is considerable research effort to replace the storage node (floating gate) by nanomaterials, to increase the device reliability. Apart from this, a 1D

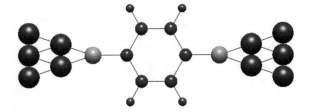

Fig. 6.7 Molecular electronics. Single molecule electronic device represents the ultimate device scaling

channel in the form of nanotubes or nanowires may enhance the electrostatic control of the channel region by gate electrode.

6.2.3 Molecular Electronics

With the continuing device scaling, the channel dimensions are approaching atomic and molecular limit. There has been ongoing research to replace silicon with other promising nanomaterials, like nanoparticles, quantum dots, nanowires, nanotubes, graphene, etc. Yet another promising area is to use atomically engineered single molecules as the channel to fabricate multi-functional devices as shown in Fig. 6.7.

Making progress in this area has been quite challenging due to unique material challenges of integrating molecules with metallic and/or semiconducting contacts. Additionally, controlling the molecules for precise deposition between the two nanoscale contacts has been tricky. However, once these bottlenecks are overcome, the rewards would be worth the effort.

For the organic molecules, chemists know how to make millions and billions with the same structure, which is instrumental in obtaining uniform device characteristics. Furthermore, most of the organic molecules have a small dielectric constant, which facilitates a better electrostatic control by the gate voltage in the channel region.

6.2.4 Thermoelectric Devices

Thermoelectric devices work on the principle of direct energy conversion of temperature difference into electrical power and vice versa. A typical thermoelectric generator is shown in Fig. 6.8a, where a hot reservoir is separated from the cold reservoir by two distinct n-type and p-type regions.

When a temperature difference is established between the two reservoirs, electrons and holes flow from the hot to the cold reservoir through n-type and p-type semiconducting material, respectively. This unidirectional flow of conventional current then delivers electrical power to a load or a battery. Such a device may be operated in power generation mode and is called a ThemoElectric Generator (TEG). Since, no moving parts are involved, this method of power generation is noise free and

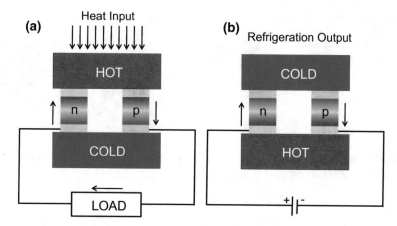

Fig. 6.8 Working principle of thermoelectric devices. **a** Electric power generation mode. **b** Refrigeration mode

relatively environment-friendly, apart from the environmental impact during device fabrication and material processing.

The same device may function in refrigeration mode by connecting an external supply voltage to one of the reservoirs as shown in Fig. 6.8b. In this case, electrons and holes are forced to travel in opposite directions to that of the power generation mode. As a result, cold side's temperature decreases and hot reservoir's temperature increases. Heat is further dissipated through the hot reservoir. Therefore, the net output is refrigeration.

Problems

6.1 For $D(E) = 1\,\text{eV}^{-1}$ and $v_g = 0$, calculate the transmission $T(E)$.

6.2 What is the value of the quantum of conductance for one band? If the applied voltage is 1 V, calculate the current.

6.3 What is the value of the quantum of conductance for two bands? If the applied voltage is 1 V, calculate the current.

6.4 Comment on the state of the art channel length in silicon transistors. How many silicon atoms are there on average along the length of the channel?

6.5 Give an example of a nanosensor.

6.6 Give an example of a nanobiosensor.

6.7 Give an example of a bionanosensor. How is it different from a nanobiosensor?

6.8 What are the various nanomaterials used in the storage node of a flash memory?

6.9 Give an example of a single molecule transistor.

6.10 What are the commonly used materials in the thermoelectric devices?

6.11 What are the advantages and disadvantages of thermoelectric refrigerators?

6.12 What are the advantages and disadvantages of thermoelectric generators?

Research Assignment

R6.1 Nanotechnology has had the most evident impact on the area of computing. This technology has revolutionized computer chips, memory chips, charge couple device (CCD) chips, sensors, flexible electronics, televisions, etc. Pick a topic of your choice about the impact of nanotechnology on computing devices, and write a one-page summary.

References

1. Desai et al., Science **354**, 99 (2016)
2. Migita et al., IEEE International Electron Devices Meeting, pp. 8.6.1–8.6.4 (2012)

Chapter 7
Spintronics

So far in our discussion of the band picture of materials, we have ignored the spin of an electron. The magnetic properties of metals like Fe, Ni, Co, etc, and magnetic semiconductors like GaMnAs, MnAs, etc. may directly be attributed to the spin, which is explored in this chapter.

7.1 Magnetic Materials

In a non-magnetic material, the bands for up-spin ($\uparrow$-spin) and down-spin ($\downarrow$-spin) electrons are degenerate and hence overlap with each other as shown in Fig. 7.1a. In this situation, the number of $\uparrow$-spin electrons and the number of $\downarrow$-spin electrons are equal. Although individual electrons have an intrinsic magnetization due to the electron spin. The overall magnetization is zero for such a band structure.

On the other hand, in a magnetic material, the bands for the two spin directions are split by what is known as the Exchange Interaction. Due to this splitting, the number of electrons with $\uparrow$-spin ($N_\uparrow$) and the number of electrons with $\downarrow$-spin ($N_\downarrow$) at the chemical potential are different. For the splitting shown in Fig. 7.1b, the number of $\uparrow$-spin electrons are higher than the $\downarrow$-spin ones, i.e. $N_\uparrow > N_\downarrow$. Due to this difference, a material like Fe achieves a net magnetization and hence manifests magnetic properties, thereby forming a ferromagnet.

These magnets may be magnetized by applying an external magnetic field. Even when the magnetic field intensity (H) is reduced to zero, there is some remanent magnetization (M_r, also called remanence) left in the sample as shown in Fig. 7.2. In order to switch the magnetization of a magnet, the remanent magnetization first needs to be reduced to zero by applying a coercive field (H_c, also known as coercivity). If the coercivity has a large value, the magnet is said to be a hard magnet, whereas for a

© Springer Nature Switzerland AG 2019
H. Raza, *Freshman Lectures on Nanotechnology*, Undergraduate Lecture Notes in Physics,
https://doi.org/10.1007/978-3-030-11733-7_7

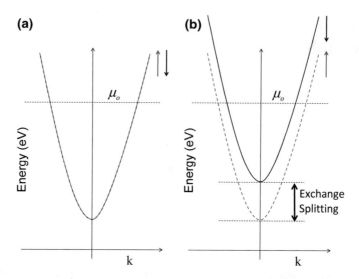

Fig. 7.1 Magnetic properties of materials. **a** Bandstructure with degenerate ↑-spin and ↓-spin bands. **b** Exchange splitting between ↑-spin and ↓-spin bands in magnetic materials gives rise to a net magnetization due to the unequal number of electrons with ↑-spin and ↓-spin around the equilibrium chemical potential (μ_o)

Fig. 7.2 Hard and soft magnet

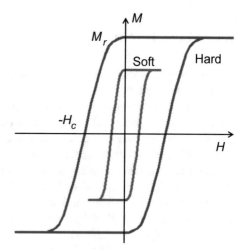

small coercivity, the magnet is termed as a soft magnet. The idea is that by applying a small magnetic field, the soft magnet may switch its magnetization without affecting the magnetization of the hard magnet. Thus, one may imagine forming a layered structure of corresponding soft and hard magnets, which we discuss next.

7.2 Giant MagnetoResistance Device

A giant magnetoresistance (GMR) device is a trilayer device, where a non-magnetic film like copper is sandwiched between two ferromagnetic films like iron, as shown in Fig. 7.3a and b.

There are two types of GMR device structures depending on the current flow direction and the interface between the various layers. For the current flow perpendicular to the interface (plane) between the layers, the device structure is called CPP (current perpendicular to the plane) as shown in Fig. 7.3a. On the other hand, for the current flow parallel to the interface (plane) between the layers, the device structure is termed as CIP (current in the plane) as shown in Fig. 7.3b.

To make the GMR device work, one should be able to selectively switch the magnetization of one of the two layers. The ferromagnetic layer that has fixed magnetization is called the fixed layer (or pinned layer), whereas the other ferromagnetic layer whose magnetization may be switched by applying an external magnetic field is called the free layer. This selective magnetization reversal is achieved by making the fixed layer out of a hard magnet and the free layer out of a soft magnet.

The device configuration where the magnetic orientations of the fixed and the free layers are in the same direction is called the Parallel (P) configuration as shown in Fig. 7.4a. On the other hand, the configuration where the magnetic orientations of the fixed and the free layers are in the opposite direction is called the Anti-parallel (AP) configuration as shown in Fig. 7.4b. The read head of a hard disk, shown in Fig. 7.4c, forms the fixed layer of the GMR device and is able to sense the magnetic orientation (corresponding to P or AP configuration) on the disk.

The current through the device in the P configuration is usually higher than the AP configuration, giving rise to a low-resistance (R_P) in the P configuration and a high-resistance (R_{AP}) in the AP configuration. Due to this bistable nature, a GMR device is used as a memory element. One figure of merit defined for these devices is the GMR ratio, given as follows,

$$\text{GMR} = \frac{R_{AP}}{R_P} - 1 \tag{7.1}$$

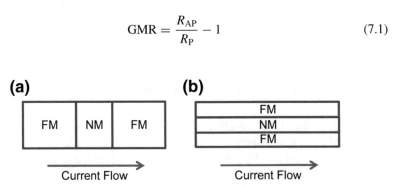

Fig. 7.3 GMR device. **a** Current parallel to plane (CPP) device structure. **b** Current in plane (CIP) device structure

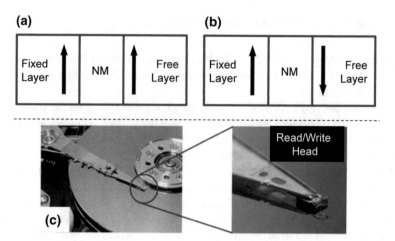

Fig. 7.4 GMR device states. **a** Parallel (P) configuration leads to a low resistance state. **b** Anti-parallel (AP) configuration leads to a high resistance state. **c** The read head of a hard disk (courtesy of IBM)

Fig. 7.5 MTJ device.
a Parallel configuration.
b Anti-parallel configuration

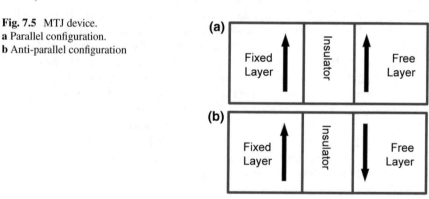

The state of the art GMR ratio is about 20% as of today.[1]

7.3   Magnetic Tunnel Junction Devices

The saturation in the GMR device research led to the development of magnetic tunnel junction (MTJ) devices, where the middle non-magnetic metallic layer is replaced by a non-magnetic insulator layer as shown in Fig. 7.5a, b for P and AP configurations, respectively. The switching between the P and the AP configurations in an MTJ device works in the same manner as that of a GMR device. However, the conduction mechanism in an MTJ device is quite different. While the conduction mechanism in

[1]2018.

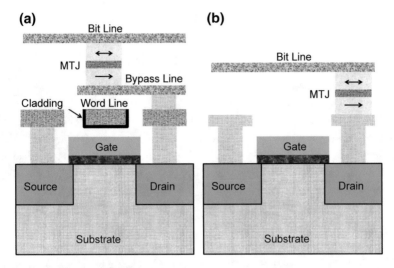

Fig. 7.6 Memory Cell. **a** MRAM. **b** STT-RAM

a GMR device is diffusive due to small de Broglie wavelength in metals, the transport in an MTJ device is based on quantum mechanical tunneling due to a thin insulating barrier separating the two ferromagnetic contacts. Based on the resistances in the P and the AP configurations, one may define TMR (tunnel magnetoresistance) ratio as follows,

$$\text{TMR} = \frac{R_{\text{AP}}}{R_{\text{P}}} - 1 \tag{7.2}$$

Initially, aluminum oxide (Al_2O_3, alumina) was used in the MTJ devices. However, the TMR ratio saturated to about 20% for the alumina based MTJ devices due to the amorphous nature of alumina. With the use of crystalline magnesium oxide (MgO), TMR ratios in excess of 1000% have been achieved. It is no surprise that the magnetic read heads in hard disks have been replaced with MTJ devices due to their superior performance.

The MTJ devices, although enjoying higher TMR ratios, still have the bottleneck that an external electromagnet is needed to switch the magnetic orientation of the free layer. This not only results in tremendous circuit complexity in the MRAM (magnetic random access memory) cell based on MTJ devices, as shown in Fig. 7.6a, but also limits the memory performance due to cross-talk between the memory cells.[2]

This shortcoming has been addressed with the development of spin transfer torque (STT) devices. In an STT device, the magnetization of the free layer is switched by using the spin polarized current flowing through the device itself, thereby eliminating

---

[2]Detailed description of an MRAM cell is left as an exercise for the curious reader.

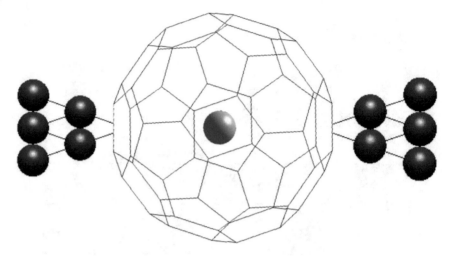

Fig. 7.7 Molecular Spintronics. Schematic of SMM-based molecular transistors. Co atom encapsulated inside a C_{60} cage

the need for an external electromagnet and associated circuitry. The associated STT-RAM cell is shown in Fig. 7.6b, highlighting the simplicity in the implementation as well as the reduction in the cross-talk between neighboring memory cells.

7.4   Molecular Spintronics

Magnetic atoms may be incorporated into caged molecules like porphyrin or Buckyball (C_{60}), and with that one steps into the regime of molecular spintronics. The motivation behind studying single molecule magnets (SMM) is not only to develop the physical understanding, but also the applied nature of their potential use in the spin based devices. With the increasing packing density for hard disks, MRAM and STT-RAM,[3] the magnets are also reaching the atomic limit. Due to their uniformity, the SMM may be an ideal candidate to replace the magnetic films on the hard disk.

As discussed in the previous chapter, just like nonmagnetic molecules are used in the molecular electronics, the SMM may be incorporated in a transistor structure to form spin transistors, where the gate voltage or magnetic field may be used to modulate the transport. Here, the molecule may be porphyrin based, e.g. [Co(TerPy)$_2$], or C_{60} based as shown in Fig. 7.7, where the Co atom imparts the magnetic behavior.

One should note that there are numerous examples of SMM found in nature. The ferritin protein that is used to transport Fe is an SMM. The hemoglobin, a blood protein, used to transport oxygen, is also an SMM.

---

[3]Detailed description of an STTRAM cell is left as an exercise for the curious reader.

Problems

7.1 What is a ferromagnetic material? Give an example.

7.2 What is an antiferromagnetic material? Give an example.

7.3 What is the superparamagnetic limit for a nanomagnet?

7.4 Give an example of a soft magnet.

7.5 Give an example of a hard magnet.

7.6 Amongst CIP and CPP configurations, which has a higher GMR ratio?

7.7 For a GMR device, $I_P = 1.05\,\mathrm{mA}$ and $I_{AP} = 1\,\mathrm{mA}$, calculate its GMR ratio.

7.8 For an MTJ device, $I_P = 3\,\mu\mathrm{A}$ and $I_{AP} = 2\,\mathrm{nA}$, calculate its TMR ratio. Consult the literature and comment on the current values in comparison with the GMR device in Problem 7.7.

7.9 What are the advantages of STTRAM? What are the bottlenecks?

7.10 Give an example of how single molecule magnets (SMM) are being used in nature or man-made structures.

Research Assignment

R7.1 Write a one-page summary about nanomagnetism and its applications.

R7.2 Communication applications like fiber optics, wireless technologies, sensors/detectors, light emitting diodes, LASER devices, have benefited from the use of nanomaterials and nanotechnology. Pick a topic of your choice about how communication technologies have benefited from the nano age, and write a one-page summary.

Chapter 8
Nanophotonics

Photonics is the study of light-matter interaction, where photons interact with electrons and vice versa. More specifically, Nanophotonics is,

the science and engineering of light matter interactions that take place on wavelength and subwavelength scales where the physical, chemical or structural nature of natural or artificial nanostructured matter controls the interactions.

—National Academy of Science

These interactions in nanomaterials at the wavelength or subwavelength scales lead to interesting phenomenon. Consider the butterfly wing and the peacock feathers shown in Fig. 8.1a, b, respectively. The *egg-crate* pattern found in the butterfly wings selectively reflects colors with very high efficiency, thereby resulting in beautiful and brilliant patterns. The case of peacock feathers is even more interesting where 2D photonic crystals found on the cortex in the peacock's feather barbules leads to various coloration.

In this chapter, we start the discussion with *photonic crystals*, which are man-made periodic structures inspired by nature that have the ability to control the flow of light.

8.1 Photonic Crystals

In free space, light propagation follows a linear dispersion, which is given as follows,

$$c = f \lambda_o \tag{8.1}$$

where c is the speed of light in free space, f is the frequency, and λ_o is the free space wavelength. When light passes through a medium, the speed of light v is given as,

H. Raza, *Freshman Lectures on Nanotechnology*, Undergraduate Lecture Notes in Physics, https://doi.org/10.1007/978-3-030-11733-7_8

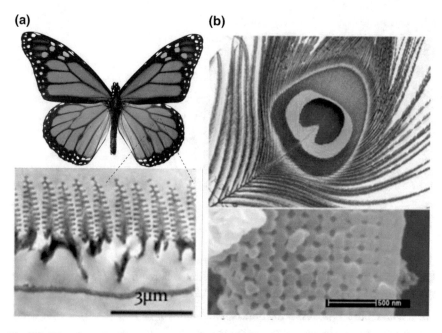

Fig. 8.1 Nanophotonics in nature. **a** Butterfly wings. **b** Peacock feathers [1, 2]

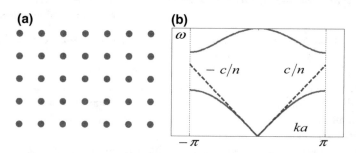

Fig. 8.2 Photonic crystal. **a** Photonic crystal with 200 nm periodicity. **b** The crystal develops nonlinear dispersion and various bands resulting in a band gap. A linear dispersion is shown as a reference

$$v = f\lambda \tag{8.2}$$

where $v < c$ and $\lambda < \lambda_o$. The ratio c/v is defined as the index of refraction (n). By using $\omega = 2\pi f$ and $k = 2\pi/\lambda$, the above linear dispersion may be rewritten as,

$$\omega = vk \tag{8.3}$$

We have studied the band gap in electronic spectrum of semiconductors earlier. Similarly, one may engineer a band gap in the optical spectrum of man-made materials by fabricating periodic structures made with materials of different indices of refraction as shown in Fig. 8.2a. Such structures are called *photonic crystals* in analogy with the semiconducting crystals.

As shown in Fig. 8.2b, the periodic structure with 200 nm period gives rise to a nonlinear dispersion and a band gap. The diffraction of light is forbidden for the range of frequencies inside the band gap. Moreover, there are two bands found in the optical spectrum of the photonic crystal. One is below the band gap and the other is above the band gap, analogous to the valence band and the conduction band in semiconductors.

The ratio of the index of refraction of the photonic crystal to that of the surrounding material is an important parameter that controls the confinement of light through the crystals and in fact various other structures. Furthermore, the band gap may depend on the direction of the light or may be engineered to be omni-directional.

8.2 Confinement (1D, 2D, 3D)

While electrons may be confined by electrostatic potentials, photons are confined by the difference in the indices of refraction between the two media. Consider the structure shown in Fig. 8.3, where the index of refraction of the inner material (n_1) is greater than that of the surrounding material (n_2). At the correct angle of incidence, most of the incident light is confined in the core through total internal reflection. The light exponentially decays in the outer material. The analogy of such a decay

Fig. 8.3 Confinement effects. $n_1 > n_2$ ensures light propagation in the inner material for a certain range of incident angles. Planar waveguide, optical fiber and microsphere have 1D, 2D and 3D confinements respectively, resulting in varying optical properties

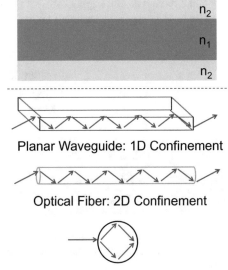

Planar Waveguide: 1D Confinement

Optical Fiber: 2D Confinement

Micro/Nano Sphere: 3D Confinement

would be that of the quantum mechanical tunneling through a barrier, where the wavefunction exponentially decays into the barrier.

As shown in Fig. 8.3, a planar waveguide leads to confinement in 1D, resulting in 2D propagation. On the other hand, an optical fiber confines light in 2D, resulting in 1D propagation along the length of the fiber. Similarly, a sphere confines light in all three dimensions resulting in light trapping. One may find analogies between the electron confinement effects in nanostructures, and the photon confinement effects in micro/nanostructure at wavelength or subwavelength scales.

Such confinement results in optical resonances, where the light absorption or emission not only depends on the frequency but also the confinement. For example, a micro/nanosphere may confine certain wavelengths of light depending on its size. A bigger sphere may be able to sustain red light, where as a smaller sphere may have resonance in blue wavelength range.

8.3   Meta Materials

So far, we have discussed materials with positive index of refraction (i.e. $+n$). Next, we discuss a new class of materials, known as meta materials, that have a negative index of refraction (i.e. $-n$).

With a $-n$, the laws of reflection and refraction lead to interesting observations. The light focusing ability of a meta material when light passes through it during refraction is shown in Fig. 8.4a in comparison with the refraction through a normal material.

In Fig. 8.4b, the refraction of light through the beaker filled with water (left) with $+n$ and a meta material (right) with $-n$ is shown. The novel refraction phenomenon may be used in engineering novel lenses. Furthermore, meta materials may be used

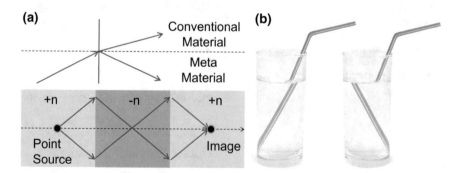

Fig. 8.4 Meta materials. **a** Laws of refraction with a negative n. A meta material internally focuses the light. **b** A visual display of the effect of meta materials on light propagation (courtesy of phys.org)

in directing light beam of certain polarization. One should note, however, that the meta materials have the desired properties only over a finite frequency range.

Although we have discussed meta materials with negative refractive index only, there are several kinds of meta materials, the discussion of which is beyond the scope of this book.

8.4 Near-Field Optics

Near-field optics deals with the interaction of light with sub-wavelength structures. It is well-known that the optical resolution is limited by the Rayleigh criterion of diffraction limit in far field, and is defined as,

$$0.61 \frac{\lambda}{NA} \tag{8.4}$$

where NA is the numerical aperture. Within the diffraction limit, there are two ways to improve the resolution. First is to increase the numerical aperture, whereas the second is to decrease the wavelength, which is achieved by immersing the optics in a liquid with high refractive index in order to reduce the wavelength by λ_o/n, where λ_o is the free space wavelength.

Yet another method to achieve better resolution is to use the near-field techniques in order to overcome the diffraction limit. In near-field optics, the diffraction limit is overcome by using nanoscale apertures or even aperture-less nanostructures to enhance the light-matter interaction at the nanoscale. The sub-wavelength apertures may be achieved by using nanoscale tips, nanospheres, etc. The rule of thumb is that the near-field decays within a distance of about 50 nm from the sub-wavelength aperture as shown in Fig. 8.5a. By using a tapered optical fiber with a 50 nm opening

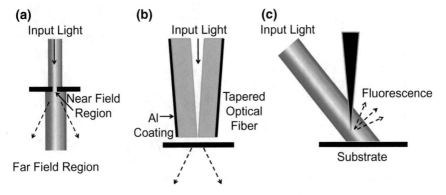

Fig. 8.5 Near-field optics. **a** Near-field versus far-field. Scattering through sub-wavelength aperture. **b** Tapered optical fiber. **c** Metal tip leading to Fluorescence

[shown in Fig. 8.5b] or by using a sharp metal tip in the proximity of the substrate [shown in Fig. 8.5c], near-field effects may be achieved. This results in sub-diffraction resolution, even with standard optical techniques. What we described is precisely the working principle of the near-field scanning optical microscope (NSOM).

8.5 Molecular Photonics

Bond between two atoms may be thought of a coupled spring vibrating at a certain frequency or in fact a set of various frequencies, leading to vibrons, which are similar to phonons in semiconducting crystals. The electric-field of these virbons couples well with the electric-field of photons. When molecules are excited with incident photons, the interactions between the vibrons and photons lead to interesting phenomenon. When a molecule absorbs a photon, e.g. of infrared (IR) frequency, it may simply be absorbed as shown in Fig. 8.6a. However, if a molecule absorbs a photon of higher energy and re-emits a photon of the same energy, this elastic process is called Rayleigh scattering as shown in Fig. 8.6a.[1] On the other hand, if the photon exchanges energy with the virbons, e.g. by emitting a vibron, the emitted light has less energy, which results in Stokes Raman scattering. On the other hand, if a vibron is absorbed, the energy of the emitted photon is higher than the incident photon, resulting in an anti-Stoke Raman scattering event. These Raman scattering features are used for characterizing nanomaterials, which we discuss further in Chap. 11.

Photonic interaction with molecules may not only be used as a probe for nanoscale characterization, but this may also be used to understand how nature works and in the process make useful nature-inspired photonic devices. One important application is the *photosynthetic solar cell* based on artificial leaf that replicates the natural process

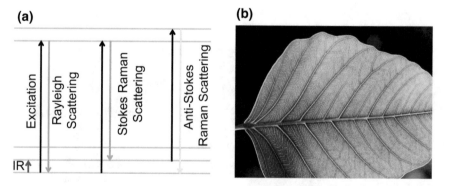

Fig. 8.6 Molecular photonics. **a** Various scattering mechanisms. **b** Learning from nature may enable novel nanophotoic devices

---

[1] This process may also be referred to as elastic scattering.

of photosynthesis as shown in Fig. 8.6b. Yet another example in nature for such mechanism is *bacteriorhodopsin*, that transports protein across a certain membrane when illuminated.

Problems

8.1 Describe the cortex structure in the peacock's feather barbules and explain how does it get its striking color pattern.

8.2 Describe the butterfly wing structure and explain how does it get its striking color pattern.

8.3 Comment on the material, spacing and photonic band gap of a photonic crystal of your choice.

8.4 What is the wavelength of the photons being used in optical fiber communication? What is the reason behind this specific wavelength? Which materials are used for making sources and detectors in optical fiber communication and why?

8.5 Comment on the cloaking application of a meta-material of your choice.

8.6 What is a super lens? How nanophotonics is helping in this regard?

8.7 For an electron microscope, $\lambda = 1\,nm$, $NA = 0.05$, what is the resolution?

8.8 How can the resolution of a microscope be improved in context of the wavelength being used?

8.9 How can the resolution of a microscope be improved in context of the numerical aperture being used?

8.10 Give an example of molecular photonics inspired application.

Research Assignment

R8.1 Energy related applications impacted by nanotechnology include solar cells, solar thermal, thermoelectric, batteries, wind energy, sensors, etc. Pick a topic of your choice about how nanotechnology is helping us cope with the ever increasing energy demand, and write a one-page summary.

References

1. Vukusic et al., Proc R Soc Lond [Biol], **266** (1999)
2. Zi et al., Proc Natl Acad Sci **100**, 12576 (2003)

Chapter 9
Nanofabrication

One may envision that nanofabrication also involves two approaches. The first is the top-down approach where the nanostructures are fabricated by patterning and etching material. The second is the bottom-up approach towards nanofabrication, which is based on how the nature works, i.e. engineering the building blocks in a way that they fit together under the right conditions.

A nanofabrication process in fact may go hand in hand with microfabrication, where the two may very well complement each other. Therefore, in a way, learning about microfabrication is imperative for nanofabrication.

In a typical microfabrication run, one starts with a substrate, usually silicon wafer. Oxidation is the first step to create a smooth and defect-free insulating film. Material deposition may be achieved by using deposition techniques like sputtering, chemical vapor deposition (CVD), epitaxial growth, molecular beam epitaxy (MBE), and physical vapor deposition (PVD), which include thermal evaporation and electron-beam evaporation. The material deposition is matched with the photolithography steps by using a mask aligner or a stepper. The excessive material is then selectively etched by using chemical etching in the form of chemicals, reactive ion etching (RIE) and inductive coupling plasma (ICP), or physical etching by using ion milling. Since various reviews of microfabrication techniques already exist, we focus on nanofabrication in this chapter.

9.1 Nanomaterial Synthesis

Nanoparticle synthesis involves methods following either the top-down approach or the bottom-up approach. In the top-down approach, the synthesis methods include attrition and milling, whereas the bottom-up synthesis techniques may be classified either as gas phase (e.g. pyrolysis, inert gas condensation, etc) or liquid phase (e.g. solvothermal reaction, sol-gel, etc).

© Springer Nature Switzerland AG 2019
H. Raza, *Freshman Lectures on Nanotechnology*, Undergraduate Lecture Notes in Physics,
https://doi.org/10.1007/978-3-030-11733-7_9

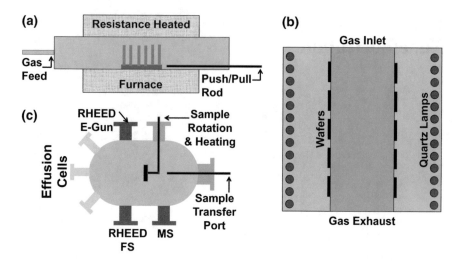

Fig. 9.1 Deposition techniques. **a** Chemical vapor deposition (CVD). **b** Epitaxial growth. **c** Molecular beam epitaxy (MBE)

The simplest of these techniques is the inert evaporate, vapor, or gas condensation, usually carried out in a PVD chamber (e.g. evaporator) or a CVD chamber. In case of an evaporator, a thin film of $1-2\,$nm is evaporated that forms nanoparticles on a substrate by self-assembly process. In case of a CVD chamber, the feed gas is introduced at high temperature in an otherwise inert helium, argon or nitrogen ambient atmosphere as shown in Fig. 9.1a. This feed gas reacts or decomposes on the surface to produce the desired nanoparticles. The gases used are usually of ultra high purity (UHP) category to ensure that the deposited materials have minimum defects and foreign atoms. Historically speaking, the discovery of Buckyball was achieved in a manner similar to the inert gas condensation. The arc produced between two graphite electrodes in an inert helium atmosphere leads to the formation of Buckyball molecules in gas phase.

CVD may also be used for depositing thin films of varying compositions. For example, the following reactions in gas phase leads to the deposition of silica and silicon nitride films, respectively,

$$2SiH_4 + 3O_2 \rightarrow 2SiO_2 + 4H_2O \qquad (9.1)$$

$$3SiH_4 + 4NH_3 \rightarrow Si_3N_4 + 12H_2 \qquad (9.2)$$

Apart from this, CVD has been successfully used to grow carbon nanotube (CNT) by using a catalyst of Fe based nanoparticles. By using appropriate size of the nanoparticle catalyst, single wall or multi wall CNTs may be grown.

Graphene may also be grown by using CVD. The starting substrate is usually copper or nickel. By using copper, monolayer graphene is grown quite reproducibly due to smaller carbon solubility in copper. On the other hand, bilayer graphene may be grown with ease on nickel substrate due to relatively larger carbon solubility in nickel. Not only this, epitaxial reactors, shown in Fig. 9.1b, may be used for growing nanoparticles and nanofilms. The working principle is the same as that of the CVD system, but usually with additional bells and whistles, like quartz lamps instead of resistive heating in CVD, etc.

Yet another technique for nanomaterial synthesis is the molecular beam epitaxy (MBE) as shown in Fig. 9.1c. The MBE chamber consists of various effusion cells containing the materials to be deposited. The sample is usually mounted on a heated stage that may be rotated. MBE chamber has an in-built thickness measurement equipment in the form of RHEED (reflective high energy electron diffraction) system. MBE has been very successfully used in growing III–V and II–VI semiconducting materials.[1]

9.2 Molecular Self Assembly

Molecular self assembly is how the nature works. It is an extremely low-power and highly efficient synthesis process, resulting in materials with excellent physical and chemical properties. Molecular self assembly is based on the notion that molecules under appropriate conditions, if engineered in a proper way, fit together as a jig-saw puzzle, with or without any external stimulus. One additional requirement of the molecular self assembly is that the process should be self-limiting.

As an example of molecular self assembly, let us consider atomic layer deposition (ALD), which has become a standard for high-quality defect-free gate deposition in MOS (metal-oxide-semiconductor) technology. ALD process for the deposition of alumina (Aluminum Oxide, Al_2O_3) is schematically shown in Fig. 9.2a–d.

The starting substrate is silicon with a monolayer of hydroxyl ($-OH$) group (only O atoms are shown). The monolayer of hydroxyl group may be formed by prior exposure to DI (de-ionized) water as shown in Fig. 9.2a. In the next step, trimethylaluminum (TMA) molecules [$(CH_3)_2Al_2$] are introduced, which react with the hydroxyl group and in the process, aluminum forms a bond with the oxygen on the surface - only Al atoms are schematically shown in Fig. 9.2b for this step. The byproduct is methane (CH_4) in this process, not shown in Fig. 9.2c. It does not matter how much TMA molecules are introduced, as far as there are enough, a monolayer of dimethylaluminum is deposited on top of the monolayer of oxygen atoms in a self-limiting process as shown in Fig. 9.2c.

In the next step, DI water molecules are introduced, the oxygen in these molecules reacts with the aluminum atoms. After enough water molecules react with the aluminum atoms, a monolayer of hydroxyl molecules ($-OH$) on top of a monolayer

---

[1] III–V materials consist of atoms from group III and group V, e.g. Gallium Arsenide, widely used for LEDs and LASER devices. II–VI materials consist of atoms from group II and group VI, e.g. Cadmium Telluride, used for night vision devices.

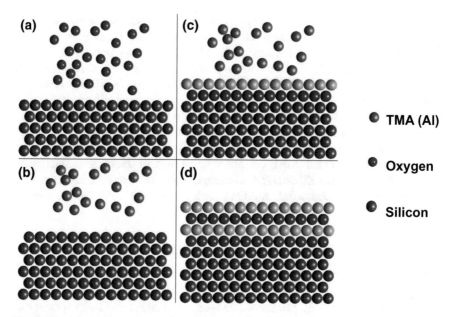

Fig. 9.2 Atomic layer deposition of alumina on silicon substrate. **a** DI (de-ionized) water exposure forming a monolayer of hydroxyl group (only O atoms shown). **b** TMA (trimethylaluminium) molecule exposure. **c** Self-limiting process. **d** Layer by layer growth. Silicon, oxygen, and aluminum atoms are shown by red, blue, and green balls

of Al atoms is formed. One may continue this cyclic process of TMA exposure and DI water exposure, to synthesize aluminum oxide layer by layer atomically in a controlled and self-limited molecular self assembly process as shown in Fig. 9.2d.

9.3 Electron-Beam Lithography

Electron-beam lithography is an example of the top-down approach towards nanofabrication. It is well-known that photolithography is a parallel process, as shown in Fig. 9.3a, where the whole mask or reticle is exposed to transfer the pattern on the substrate by using an ultraviolet (UV) source. In contrast, the electron-beam lithography (EBL) is a serial process, which we discuss next.

In electron-beam lithography, a sharp electron beam with the size on the order of a few nanometers is formed in an electron optics column. Electromagnetic lenses are used to focus the beam emanating from the electron gun assembly on the substrate. More electromagnetic lenses are used as beam deflectors, which help scan the substrate that contains the electron-bream resist, usually in the form of PMMA [Poly(methyl methacrylate)]. After development, the electron-beam resist

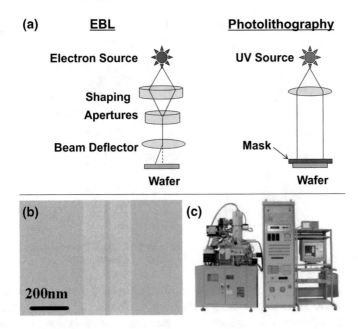

Fig. 9.3 Electron-beam Lithography (EBL). **a** Serial versus parallel nature of EBL and photolithography. **b** Fabricated structures by using EBL. **c** A commercially available electron-beam writer (courtesy of JEOL)

gets removed from the exposed areas, leaving behind nanosacle features, which may be etched or deposited on, depending on the process run.

A fabricated structure by using electron-beam lithography is shown in Fig. 9.3b, where the line-widths of gold features are about 130 nm each and the spacing between the lines is about 20 nm. In the state of the art electron-beam lithography systems, shown in Fig. 9.3c, resolution of up to a few nanometers may be achieved. However, low throughput is an inherent disadvantage of the system due to the serial nature of the process and the subsequent slow speed. In this context, there have been some efforts to parallelize the electron-beam lithography systems by employing multiple electron-beams.

9.4 Nanoimprint Lithography

Nanoimprint lithography is a low cost, high resolution, and high throughput nanofabrication technique. It has become quite popular recently due to the feasibility of its integration with photolithography equipment. However, the mold has a finite lifetime, a bottleneck due to the contact nature of the imprinting process.

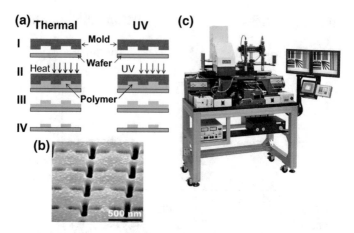

Fig. 9.4 Nanoimprint lithography. **a** Comparison of thermal and UV nanoimprint lithography. **b** Fabricated structures. **c** Commercially available system. (courtesy of OAI)

The starting point of the nanoimprint lithography is the mold containing the nanoscale features as shown in Fig. 9.4a [step I]. The mold is usually fabricated by using electron-beam lithography technique. There are two variants of nanoimprint lithography, namely thermal and UV (ultraviolet). In the thermal nanoimprint lithography, a thermoplastic polymer is spun coated on the substrate. The mold is subsequently pressed against the heated thermoplastic polymer film containing the substrate [step II]. Heating is critical to improve viscosity. The whole assembly is then cooled, which solidifies the thermoplastic polymer. After de-embossing the mold [step III], a negative imprint of the mold is left in the thermoplastic. An etching step [step IV] removes the excess thermoplastic thus transferring an exact negative of the mold on the substrate.

The UV nanoimprint lithography system is exactly the same as that of the thermal nanoimprint lithography system, except that a UV sensitive polymer resist is used [step II], which may be cured by UV light as shown in Fig. 9.4a. During the curing process, the resist solidifies, thereby transferring the negative of the mold on the substrate. Due to a heat-less process and availability of the UV source in photolithography systems, UV nanoimprint has become the method of choice.

A fabricated device structure by using nanoimprint lithography is shown in Fig. 9.4b. A commercially available photolithography system with nanoimprint lithography add-on is shown in Fig. 9.4c.

9.5 SPM Based Nanofabrication

Scanning probe microscopy (SPM) is a highly effective microscopy technique which is widely used for imaging nanostructures with sub-nm resolution. While we discuss the microscopy and spectroscopy applications of SPM in the next chapter, we would like to discuss the nanofabrication aspects of SPM here.

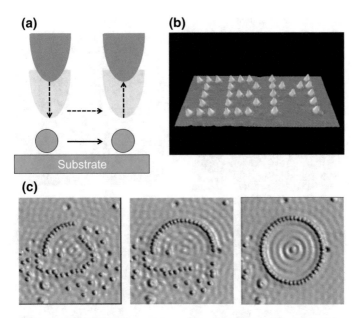

Fig. 9.5 STM based nanofabrication. **a** Manipulation of atoms to create nanostructures. **b** IBM written with individual atoms. **c** Quantum mirages formed by STM nanofabrication [1]. (courtesy of IBM)

An SPM consists of a probe that interacts with the substrate in a controlled manner, which is ensured by a feedback loop. There are predominantly two types of SPM depending on the tip and the substrate. The first is called STM (scanning tunneling microscope), whose working principle is based on the quantum mechanical tunneling through a small gap (usually in ultra high vacuum) between the tip and the substrate. Due to this, STM may only be used for conducting substrates by using tungsten or Pt/Ir tips, which are conducting as well. The detailed working principle will be discussed in the next chapter.

By bringing the tip in close proximity to the atom on a metallic substrate, the tunnel gap is reduced, resulting in a tunnel current. While maintaining the tunnel current, the atom may be moved around with the tip motion. When the atom is placed at its intended position, the tip is retracted. This increases the tunnel gap between the tip and the substrate resulting in reduced tunnel current as shown in Fig. 9.5a. Back in 1989, IBM reported the STM image of 35 xenon atoms arranged on nickel surface as shown in Fig. 9.5b. A circular atomic pattern with quantum mirage is shown in Fig. 9.5c. Various stages of the circular arrangement of atoms during the atomic manipulation are also shown.

The second type of SPM is AFM (atomic force microscope), which was developed to image insulating substrates with high resolution. In AFM, a tip is scanned over the substrate. The tip-sample interaction forms the working principle of AFM, which we discuss in the next chapter.

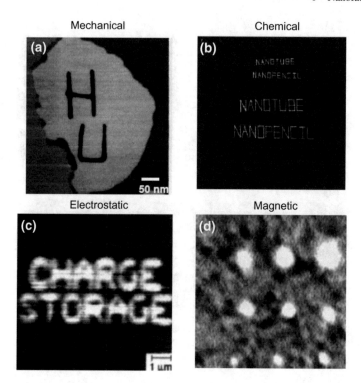

Fig. 9.6 Atomic force microscopy for nanofabrication. **a** Mechanical. **b** Chemical. **c** Electrical. **d** Magnetic [2–4] (courtesy of nanoscience.de)

In AFM, there are various kinds of tip sample interactions that may be put to use for fabricating nanostructures. Here, we discuss mechanical, chemical, electrical, and magnetic interactions as shown in Fig. 9.6a–d, respectively. One may physically scratch the substrate surface to *draw* lines at the nanoscale. In addition, the tip may be able to oxidize the substrate by applying a tip voltage in a humid environment. In case of nanomaterials that may act as charge storage nodes on the substrate, a conducting tip may electrostatically charge the substrate, which may be detected by using electrostatic attraction or repulsion between the tip and the substrate. A magnetized AFM tip may act as a read/write head, similar to the read head of a hard disk. Magnetic information may be stored in the nanomagnets on the substrate. Dip Pen lithography is yet another AFM based nanofabrication technique for depositing organic molecules on substrates. In dip pen lithography, by controlling the tip motion and the chemical environment, nanoscale features may be transferred to the substrate.

Problems

9.1 Give an example of PVD method for nanoparticle/nanocrystal growth.

9.2 Give an example of PVD method for nanowire growth.

9.3 Give an example of PVD method for nanofilm growth.

9.4 Give an example of CVD method for nanoparticle/nanocrystal growth.

9.5 Give an example of CVD method for nanowire growth.

9.6 Give an example of CVD method for nanotube growth.

9.7 Give an example of CVD method for nanofilm growth.

9.8 Which method is used for growing materials for night vision applications?

9.9 Give an example of molecular self-assembly process not discussed in this chapter for growing nano particles or quantum dots.

9.10 Give an example of molecular self-assembly process not discussed in this chapter for growing nanofilms.

9.11 What is the state of the art resolution in electron-beam lithography.

9.12 What is the state of the art resolution in nanoimprint lithography.

9.13 What is the state of the art resolution in SPM (AFM and STM) based lithography.

Research Assignment

R9.1 Defense applications of nanotechnology include materials (armor, fuselage), bullet-proof textiles, sensors (physical, chemical, biological), devices (MEMS, NEMS, Micro/Nano Electromechanical Systems), actuators, lab-on-chip, guidance and control, power systems like batteries, intelligent and smart systems, soldier support system, etc. Pick a topic of your choice about the role of nanotechnology in contemporary defense industry, and write a one-page summary.

References

1. Crommie et al., Science **262**, 218 (1993)
2. Kim et al., Science **257**, 375 (1992)
3. Matsuda et al., Appl. Phys. Lett. **78**, 1508 (2001)
4. Barrett et al., J Appl. Phys. **70**, 2725 (1991)

Chapter 10
Nanocharacterization

In this chapter, we discuss various microscopy and spectroscopy techniques used for the characterization of nanomaterials. Microscopy is indeed the science of visualizing materials experimentally, whereas spectroscopy is concerned with the metrology (the science of measurement) of spectra produced during the interaction between the matter and the external stimuli, e.g. applied bias, electromagnetic radiation, etc.

To give an idea of the length scale and the associated microscopy techniques, various examples are shown in Fig. 10.1 that are applicable at respective length scales. With a naked eye, one may be able to see up to 0.1–1 mm. To visualize cells and bacteria up to 0.1–1 μm, one requires optical or light microscopes.

Optical microscope makes the use of a light source, where the eye or a CCD (charge couple device) sensor is used as the light detector. Since the optical wavelength range is 0.4–0.7 μm, the diffraction limited resolution is in sub-micron range, depending on the numerical aperture. However, special models of optical microscopes are applied frequently to observe nanomaterials as well. For example, dark-field optical microscopy is used to detect carbon nanotubes. One may also be able to optically distinguish between monolayer graphene, bilayer graphene, and so on, if the specimen is placed on a 300 nm SiO_2 substrate.

To probe at an even smaller length scale, electron microscopes are used. In electron microscopes, electrons are used for microscopy with the wavelength tuned by the acceleration voltage, and focusing achieved by electromagnetic lenses. In transmission electron microscope (TEM), the image is formed by the transmitted electrons through the specimen. In scanning electron microscope (SEM), reflected electrons or secondary electrons are used for imaging. As discussed in the previous chapter, yet another type of microscope that makes use of the quantum mechanical tunneling of electrons across a vacuum gap is called scanning tunneling microscope (STM).

On the other hand, one may identify nanomaterials due to their chemical signature by using various spectroscopy techniques. If one uses UV (ultraviolet) and visible radiation, the spectroscopy is called UV-Vis. By using this technique, one may probe the optical and UV properties of materials. By using infrared (IR) radiation, one may probe various vibrational modes in nanomaterials. Finally, inelastic processes in materials may be studied by using Raman spectroscopy. Various fea-

© Springer Nature Switzerland AG 2019
H. Raza, *Freshman Lectures on Nanotechnology*, Undergraduate Lecture Notes in Physics,
https://doi.org/10.1007/978-3-030-11733-7_10

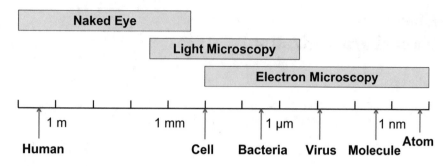

Fig. 10.1 Wide ranging length scales and associated microscopy techniques

tures in Raman spectra may be used to extract structural information about carbon nanotubes, graphene, etc. We elaborate on all these techniques in this chapter.

10.1 Electron Microscopy

In electron microscopes, the wave properties of electrons are used for imaging nano-materials, nanostructures and nanodevices. Since the optical range of wavelength is in microns, one may use optical microscope for micron size imaging. In an electron microscope, the electrons are accelerated to high velocities by using voltages in kV range. This helps to reduce the electron wavelengths to sub-nm regime, which enables nanoscale resolution by making use of electron optics.

The ray diagrams are shown in Fig. 10.2 for scanning electron microscope (SEM) and transmission electron microscope (TEM), in comparison with the optical micro-

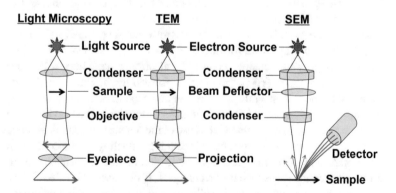

Fig. 10.2 Ray diagrams showing light microscopy in comparison with transmission electron microscopy (TEM), and scanning electron microscopy (SEM)

scope.[1] Electrons are accelerated by using an applied bias V, the kinetic energy acquired is given as:

$$E_K = q|V| \tag{10.1}$$

Equating with (2.8), one may calculate speed v in non-relativistic limits as:

$$v = \sqrt{\frac{2q|V|}{m_e}} \tag{10.2}$$

where q is the charge of electron and m_e is the electron mass.

Given the bias-dependent speed v, one may calculate the de Broglie wavelength by using (2.30). The accelerating voltage range is usually in 1–100 kV range. With a voltage of 1 kV, the wavelength is 0.04 nm or 0.4 Å. A friendly reminder that the interatomic distance is usually in [1 Å, 5 Å] range, where $1 \text{ Å} = 10^{-10} \text{ m} = 0.1 \text{ nm}$. Hence, one may use these electron waves to obtain atomic resolution. The electron waves are further focused by using electromagnets. The electrons may be accelerated close to the speed of light in these microscopes, which require relativistic corrections, the discussion of which is beyond the scope of this book.

When light is incident on a surface, part of it gets reflected, some gets absorbed and the rest gets transmitted. Similarly, when an electron wave is incident on a specimen, part of it gets bounced back. The study of these electrons gives rise to a technique called scanning electron microscopy (SEM). A commercially available SEM is shown in Fig. 10.3a. The working principle of SEM is based on the notion that the electron beam is scanned in a raster fashion across the specimen and focused to few nm or so to enable high resolution. The resulting SEM viewgraph for a CNT specimen in shown in Fig. 10.3b. Yet another advantage of electron microscopes is that these have a large depth of focus, and hence are widely used for imaging biomaterials at micron scale or even mm scale. Consider the SEM viewgraph of a mite shown in Fig. 10.3c with an excellent focus over few mm. Having such a depth of focus is a feat impossible to achieve with an optical microscope.

With an appropriate sample preparation, e.g. by thinning the specimen to 100 nm or so, the high energy incident electrons simply transmit through the specimen. This feature is used in transmission electron microscopy (TEM), as shown in Fig. 10.4a. The electron beam strikes the sample over a wide area, which produces the TEM viewgraph. A cross-section of Al–SiO_2–Si obtained by using TEM as shown in Fig. 10.4b. One may see individual Si atoms in the single crystal substrate, whereas there is no atomic order in SiO_2 due to the amorphous nature of this film. Small range order in Al is observed, which is a characteristic of the polycrystalline materials. Usually better resolution is achieved by using TEM as compared to SEM. To take advantage of both SEM and TEM, there is a hybrid between SEM and TEM, called STEM (scanning transmission electron microscopy), the discussion of which is beyond the scope of this book.[2]

---

[1]Detailed understanding of these microscopes are left as a exercise for the reader.
[2]Interested readers are encouraged to explore STEM further.

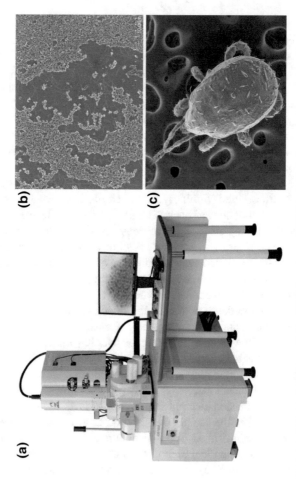

Fig. 10.3 Electron microscopy. **a** A commercially available SEM (courtesy of JEOL). **b** Carbon nanotube sample with the catalyst used for the synthesis, imaged by using SEM. **c** SEM viewgraph of a mite showing the depth of focus (courtesy of Hitachi)

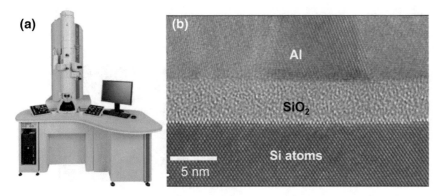

Fig. 10.4 TEM. **a** A commercially available TEM (courtesy of JEOL). **b** A TEM viewgraph showing Al–SiO$_2$–Si interface with atomic resolution

10.2 Scanning Probe Microscopy

Scanning probe microscopy (SPM) encompasses a set of techniques where a probe is used to image the sample with nanoscale resolution. Depending on the tip-sample interactions, there are various types of SPM techniques.

If both the tip and the sample are conducting, one may apply bias across a tunnel gap on the order of a nm or so between the tip and the sample, and observe what is known as tunnel current (I) as shown in Fig. 10.5a. This simple arrangement and the physical principle is used in the design and operation of a microscope, called scanning tunneling microscope (STM). We have discussed the use of STM in lithography in the previous chapter. Here, we highlight its use as an atomic scale microscope.

Since the tunnel current exponentially depends on the tunnel barrier width, about 90% of the current flows through the tip atom. One may visualize the tip atom to be

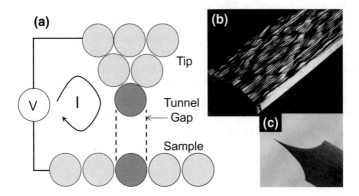

Fig. 10.5 STM. **a** Working principle. **b** Si surface visualized by using STM. **c** STM tip [1]

(a)

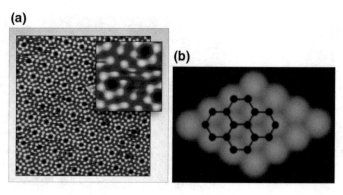

(b)

Fig. 10.6 STM viewgraphs of Si(111) surface (courtesy of RHK Tech) and graphene (courtesy of Omicron)

in between two atoms on the substrate, which increases the effective gap between the tip atom and the sample surface resulting in exponentially smaller current. When the tip atom aligns with an atom on sample surface, exponentially higher current flows as compared to when the tip atom is misaligned with the sample surface atoms. By keeping track of this tunnel current difference or modulation, one may obtain atomic scale resolution.

Atomic resolution obtained on a silicon sample is shown in Fig. 10.5b - this viewgraph is one of the first few STM images. An STM tip is also shown as a reference in Fig. 10.5c. Furthermore, an STM viewgraph of a Si(111) surface with (7×7) reconstruction[3] is shown in Fig. 10.6a. An STM viewgraph of graphene is shown in Fig. 10.6b with overlaying graphene structure emphasizing that STM really probes the electron orbits around the atoms, and not the atomic cores themselves.

STM may only be used for conducting samples and tips, since the working principle is based on the conduction through the tip-gap-sample structure. To fill the void for insulating substrates, atomic force microscopy (AFM) was developed. The working principle is based on how a blind person feels the surface of an object. An AFM tip works the same way. By bringing the tip (mounted on a cantilever) close to the surface, one enables the tip-sample interaction that bends or twists the tip as shown in Fig. 10.7a. There may be different types of tip sample interactions depending on the nature of the tip and the sample. The simplest of which is the Van der Waal forces. For a magnetic tip and a magnetic substrate, magnetic force is the dominant interaction. Electrostatic interaction dominates charged substrates and tips. In addition, LASER beam is used to detect the tip motion with good precision. Consider the AFM viewgraph of Si(111) surface with (7×7) reconstruction, where the atomic resolution is achieved as shown in Fig. 10.7b.

---

[3] A (7×7) reconstruction means that the unit cell is seven times larger on the surface than the bulk unit cell.

Fig. 10.7 AFM. **a** Working principle. **b** UHV AFM viewgraph of Si(111) surface (courtesy of Unisoku). **c** Various operation modes of AFM

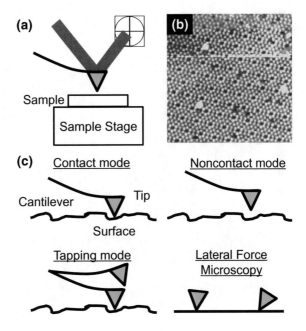

Apart from this, there are various modes of AFM operation as shown in Fig. 10.7c. In contact mode, the AFM tip physically touches the sample. In non-contact mode, the tip maintains a small distance from the sample, whereas in intermittent contact (or tapping) mode, the tip maintains a gap with occasional tapping on the sample surface. Finally, in the lateral force mode, the lateral force on the tip is tracked, which is highly effective in studying friction at nanoscale - a topic very much relevant to tribology.

10.3   Spectroscopy

The working principle of any spectroscopy technique is based on exposing the sample to the electromagnetic radiation of known wavelength or frequency range. The response of the sample to the radiation is subsequently measured.

A nanomaterial may absorb, reflect, elastically or inelastically scatter the electromagnetic radiation, depending on the composition and the arrangement of atoms and molecules constituting the nanomaterial. What makes spectroscopy a very powerful technique is the information carried by the signal which contains the chemical signature of the nanomaterial(s) under study.

The electromagnetic radiation used in most spectroscopy techniques fall in the IR (infrared), Vis (visible), and UV (ultraviolet) range. Most nanomaterials have active properties in this wavelength range.

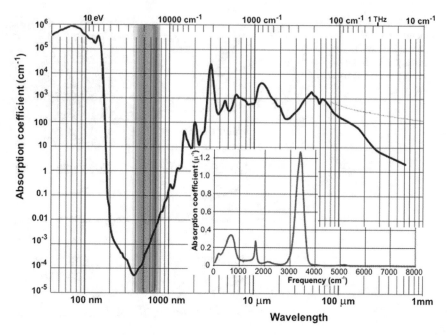

Fig. 10.8 UV-Vis spectroscopy of water (courtesy of OMLC). Absorption coefficient (in the units of cm^{-1}) is shown as a function of the wavelength (bottom axis) and the wavenumber (top axis)

In the UV-Vis spectroscopy, the sample is exposed to electromagnetic radiation in the ultra violet (UV) and visible range. The absorption is measured as a function of the wavelength and the wavenumber.[4] A sample UV-Vis spectrum of water is shown in Fig. 10.8, which has a very small absorption coefficient in the visible range, and has high absorption coefficient in the UV range.

The modern version of IR spectroscopy is based on the Fourier Transform (FT) technique as shown in Fig. 10.9a. The IR source output (with a wavenumber range of 400–4000 cm^{-1}) is fed to an interferometer, which generates an interferogram. The sample is exposed to the input IR interferogram, the output of which is detected by using an IR sensor. The output interferogram is sent to a computer, which analyzes the output by performing Fourier analysis. One should note that not all molecular vibrational modes are IR active. Various stretching, bending, deformation, rocking and bending modes of the bonds give rise to characteristic absorption features, which show up as dips in the transmittance spectrum as shown in the FTIR spectrum of polystyrene in Fig. 10.9b.

Now a days, the sensitivity of the FTIR spectroscopy has been enhanced to a point where the features in the transmittance may be directly associated with the specific nanomaterials and hence may be used for identification or bond orientations. For

---

[4]In spectroscopy literature, wavenumber is simply the inverse of the wavelength. One should note that this definition has a factor of 2π missing based on the definition in Chap. 2.

Fig. 10.9 FTIR. **a** Working principle. **b** Polystyrene spectrum (courtesy of ThermoFisher)

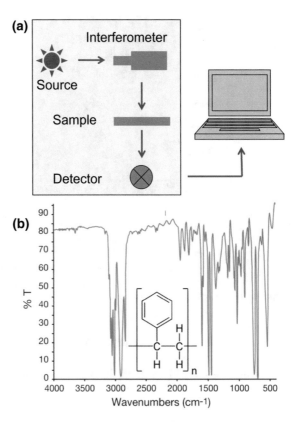

example, the specific orientation of the Si–H bond on the silicon surface may be detected by polarized IR radiation oriented along the bond length.

While discussing molecular photonics in Chap. 8, we studied various scattering mechanisms, like Raman scattering, Rayleigh scattering, etc. Raman scattering is inherently an inelastic process where the incident photons (usually from a LASER source) exchange energy with the molecular vibrations. However, not all vibrational modes are Raman active.

We report the Raman spectrum of a carbon nanotube obtained with a 488 nm wavelength LASER source in Fig. 10.10. One should note that the horizontal axis is the Raman shift (in wavenumbers, cm^{-1}) due to inelastic scattering. Various peaks are the characteristic features of the carbon nanotubes. For example, the dominant peak at $1582.4\,cm^{-1}$ is called the G peak, and is due to the in-plane vibration of the carbon lattice crystal. The peak at $1346\,cm^{-1}$ is called the D peak and is due to the defects present in the material. Finally the peak at $157.78\,cm^{-1}$ is due to the radial breathing mode (RBM) of the carbon nanotube. It is interesting to note that one may be able to calculate the diameter of the nanotube based on the wavenumber (given in cm^{-1}) of the radial breathing mode (ν_{RBM}), and is given as,

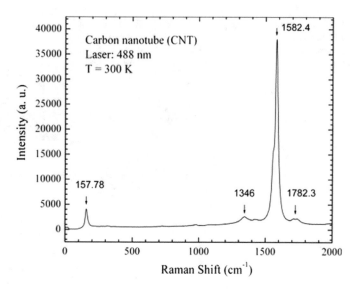

Fig. 10.10 Raman spectrum of a carbon nanotube (courtesy of ThermoFisher)

$$d(nm) = 248/v_{RBM}(cm^{-1}) \qquad (10.3)$$

For the $v_{RBM} = 157.78\,cm^{-1}$, the diameter is about 1.57 nm.

Problems

10.1 What is the minimum resolution obtained with unaided eye?

10.2 What is the minimum resolution obtained with light microscopy?

10.3 What is the minimum resolution obtained with electron microscopy?

10.4 For an electron microscope with 1 KeV accelerating voltage and $NA = 0.02$, what is the minimum resolution?

10.5 What are the advantages and disadvantages of SEM, TEM, and STEM?

10.6 Why STM does not work very well in air?

10.7 What are the advantages and disadvantages of various modes in AFM?

10.8 What are the typical applications of UV-Vis spectroscopy? What materials are usually characterized by using this technique?

10.9 What are the typical applications of FTIR spectroscopy? What materials are usually characterized by using this technique?

10.10 For the $v_{RBM} = 112\,cm^{-1}$, calculate the single-wall carbon nanotube diameter.

Research Assignment

R10.1 Food related applications of nanotechnology include sensors, tracers, packaging, food protection, condition and abuse monitors, contaminant sensors, reaction engineering and heat transfer, molecular synthesis, nanoparticles in food, water purification, equipment sanitation, etc. Pick a topic of your choice about how nanotechnology is playing a role in the food industry, and write a one-page summary.

Reference

1. Binnig et al., Phys. Rev. Lett. **50**, 120 (1983)

Chapter 11
Safety, Health, Environmental and Societal Impact

In this chapter, we discuss the safety, health, and environmental impact of Nanotechnology. We also explore how the novel applications enabled by nanotechnology may impact the societal values, ethics and sustainability.

Let us begin with the question that *why the safety, health and environmental effects of nanomaterials are of great concern?* We have discussed earlier that the material properties change altogether at the nanometer scale. In a hypothetical experiment, a 10 mm diameter gold ring appears green. In fact, it appears red if one further reduces the diameter to about 1 nm. The electrical, mechanical, chemical, thermal, optical and plasmonic properties of a material change altogether at the nanoscale. Furthermore, the atomic arrangement at this scale dictates the physical and chemical properties. Consider diamond, one of the hardest and the most expensive materials[1] known to human beings, which essentially consists of carbon atoms arranged in a specific order at the nanoscale. The same carbon atoms arranged in a layered structure leads to graphite, which is one of the softest and the cheapest material known to us.[2]

These widely varying physical and chemical properties influence the potential toxicity. In this context, evaluating both the short-term and the long-term effects of nanomaterials on organ systems and human tissues need to be extensively studied. Determining biological mechanisms for potential toxic effects, and creating and integrating models to assist in assessing possible hazards is also an important area of study. The laboratory tests and clinical trials ought to be carried out to study the possible safety, health and environmental effects on animals and eventually humans.

---

[1] As of 2018.

[2] Graphite is used in *lead pencils* and not lead itself – a popular misconception.

© Springer Nature Switzerland AG 2019
H. Raza, *Freshman Lectures on Nanotechnology*, Undergraduate Lecture Notes in Physics,
https://doi.org/10.1007/978-3-030-11733-7_11

Fig. 11.1 Personal
protection equipment
(courtesy of OSHA)

11.1 Safety

Personal protection equipment (PPE) is a vital component of the safety aspects of
Nanotechnology. Primarily, there are three routes of exposure:
(1) Inhalation, which affects the respiratory system.
(2) Dermal, where the exposure is through skin, body openings or open wounds.
(3) Ingestion, where the exposure is through eating or drinking.

The PPE should be able to limit and ideally eliminate exposure through all three
routes. We provide a viewgraph of a typical PPE used in a nanotechnology laboratory
in Fig. 11.1.

Eye and face protection is quite important in dealing with nanomaterials. Respiratory protection against inhalation of harmful nanomaterials is also relevant. To
take care of the dermal exposure, hand protection should also be considered. In this
context, one should worry about the proper thickness and appropriate material for
the gloves.

Once a hazard has been discovered and handling rules established, it is vital to
communicate the hazard and its details to the concerned stakeholders. Occupational
exposure limits (OEL) should be well-advertised and communicated to the users and
the handlers of nanomaterials. Although OEL of various nanomaterials may not be
known for more recent nanomaterials, it is recommended to consider general duty
clause of OSHA (occupational safety and health administration, USA), or equivalent
in other countries or regions.

11.2 Health

Inhalation of harmful nanomaterials generally leads to pulmonary diseases. A word of caution that the nanophases of otherwise inert materials may in fact be quite toxic. Consider PTFE (commercially known as Teflon), a material considered so safe that it is even used in pots and pans for cooking and baking. However, it has been reported that PTFE nanoparticles of 20 nm diameter could be fatal to rats at an airborne concentration of about $50 \, g/m^3$.

The dermal exposure of nanomaterials could lead to various diseases as well. Direct penetration of particles of sizes on the order or 1000 nm through skin have been reported. Such penetration is based on the diffusion process and in fact may be more pronounced for smaller particles. In this context, the penetration of nanoparticles through skin is likely and presumably highly toxic.

The third and the final exposure route is ingestion. Various diseases due to hand, mouth and food contamination have been reported for chemical exposure. Since nanoparticles have extremely large surface area to volume ratio, the surface reactivity is quite high as compared to the bulk phase. This increased surface reactivity may have extremely toxic effects on various organs and tissues.

We show TEM images of the lung of a rat with inhaled titanium oxide (titania, TiO_2) nanoparticles in Fig. 11.2. The titania nanoparticles agglomerate in macrophase are shown in Fig. 11.2a. Traces of nanoparticle exposure are also found on the lung surface as shown in Fig. 11.2b. The significance of this study lies in the fact that lungs are usually able to expel most of the foreign material, however, nanomaterials clearly fall in a different category. Since our bodies have not been evolved to handle these man-made materials. Hence, they usually end up causing highly toxic effects. The take home message is that the exposure to nanomaterial may lead to toxic effects, therefore extreme caution must be taken while working with these materials.

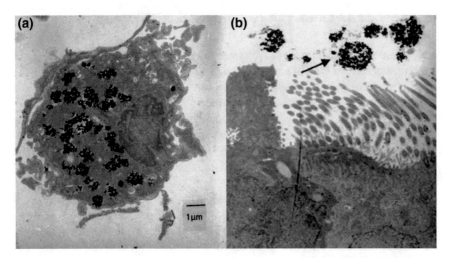

Fig. 11.2 TEM images of inhaled titania nanoparticles, **a** inside lungs, and **b** on the lung surface of a rat [1]

Table 11.1 Environmental exposure in different phases

Synthesis phase	Synthesis medium	Inhalation risk	Dermal risk	Ingestion risk
Gas	Air	Airborne toxicity	Product handling	Product leakage
Vapor	Substrate	Vapor inhalation	Vapor exposure	Chemical exposure
Colloidal	Suspension	Deposition in lungs	Product spillage	Contamination

11.3   Environment

Let us now discuss the environmental impact of nanomaterials. Various synthesis processes, e.g. gas phase, vapor phase, and colloidal, bear inhalation, dermal and ingestion exposure risks as shown in Table 11.1. The gas phase synthesis involves nanoparticle formation in air. This process may leak nanoparticles directly from the reactor during product recovery phase, or during post recovery processing and packaging. While the vapor phase synthesis process leads to nanoparticle formation on a substrate, the inhalation risks are the same as that of the gas phase synthesis.

In colloidal phase synthesis, the particle formation is in liquid suspension with potential inhalation risks during product synthesis and processing. The potential dermal and ingestion risks for the various synthesis processes entail airborne contamination of the workplace, dry contamination of the workplace, product handling, plant cleaning/maintenance, and spillage. Once a nanomaterial is released in the environment, it becomes a part of the eco-cycle and food chain, resulting in a toxic exposure to various species. In short, while dealing with a nanomaterial, one should have a complete picture of its life-cycle.

11.4   Societal Values, Ethics and Sustainability

While there are various facets of nanotechnology, which are already impacting the societal values, one aspect of nanotechnology is quite evident and is already playing its role in everyday life. Consider the area of computation that has evolved from bulky vacuum tubes to micro- and nano-chips. With the ever-increasing number of nanotransistors on the chip, the computing power in the present day devices is astounding. This development has led to widespread use of cyberspace and virtual environment. The role of digital domain, the freedom of expression and the associated privacy issues are important contemporary issues having profound impact on societal values.

Apart from this, the psychological issues like the use of the nanotechnology products (e.g. computing devices) in the development of social skills of the younger generation are also debatable. On the other hand, the role of advanced drug delivery and nanomedicine on the behavioral aspects of masses in health and food choices would be consequential.

Ethics is also expected to play a major role in the development and dissemination of nanotechnology. While the legal entities would establish the laws based on the contemporary culture and requirements, the ethical dilemmas indeed follow as a result of the lawful actions. In other words, what is lawful may not be ethical and vice versa. It would be interesting to see how ethics play a role during the evolution and development of nanotechnology in the next few decades.

Due to the scope of nanotechnology, all segments of the contemporary society are involved in the development and dissemination of this novel technology. The consumers of the nanoproducts are already enjoying various conveniences. Scientific and teaching communities have embraced the novel applications in order to exploit their potential to the fullest. The technological development has led to new industries, for which a public-private partnership has been vital. Governmental agencies further have a role to play in providing regulatory and oversight activities. Law makers have to be involved for providing legal frameworks for the use of nanotechnologies related to the environment, health, food, etc.

Apart from this, the emerging economies have an opportunity as well as a role to play in the future developments of nanotechnology. At this point, one has a level-playing field for various economies. If the emerging economies decide to invest in the development of their nanotechnology sector now, there is little doubt that their economies would flourish due to knowledge-based activities in the nano era. This will greatly help to reduce the gap between the developed and the developing (or underdeveloped) countries, resulting in variance reduction associated with global poverty. In the context of nanotechnology, the variance reduction may only be achieved through the use of economic actions (e.g. taking economic decisions at the right time, etc), as well as regulatory actions (e.g. providing tax incentives to nanotechnology industries, etc).

On the other hand, nanotechnology is in its infancy for most of the applications as of now,[3] which puts this technology in the Stage I of the Law of Diminishing Returns as shown in Fig. 11.3, where the units of output are increasing with the units of input. Hence, there is a huge opportunity for the economies to make strides in this emerging market. One should note that the different aspects of nanotechnology are expected to enter Stage II at different times. Consider the computational aspect of nanotechnology, which seems to be already in stage II due to market saturation as a result of severe competition spanning over continents. The likes of Intel, Samsung, TSMC (Taiwan Semiconductor Manufacturing Corporation), and Global Foundries compete on global scale to capture their market shares.

Most nanotechnology companies are expected to go through the three stages of the law of diminishing returns during their lifetime. During the initiation phase, the startup companies would be in Stage I. After some time, maturity of sales would put the companies in Stage II. Finally, when the sales start declining, the companies would experience the Stage III.

Just like any new technology, the societal impact must be well-studied while the technology is evolving. We cannot emphasize this enough for Nanotechnology.

---

[3] As of 2018.

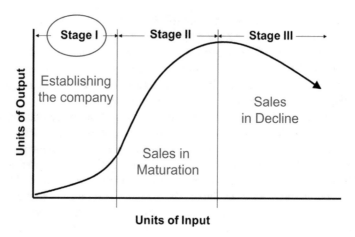

Fig. 11.3 Law of diminishing returns

Safety, Health, Environmental, and Societal Impact should be carefully evaluated. The prospects and challenges of this technology are huge and so is its impact. The rules of sustainability and ethics should be strongly tied with the advancement and development of this technology.

Problems

11.1 What are the OSHA PPE requirements for nanomaterials? Visit OSHA website for the information. You may find requirements of an equivalent organization.

11.2 What are the recommendations for glove material and thickness for handling nanomaterial(s) of your choice?

11.3 What are the recommendations for glove material and thickness for handling hydrofluoric (HF) acid.

11.4 What is Kyoto protocol. How does it affect chip fabrication industry? If the protocol is not valid anymore, please answer the question according to the protocol/treaty/agreement that replaces Kyoto protocol.

11.5 Give an example of inhalation exposure of nanomaterials. What are the potential toxic effects?

11.6 Give an example of dermal exposure of nanomaterials. What are the potential toxic effects?

11.7 Give an example of ingestion exposure of nanomaterials. What are the potential toxic effects?

11.8 Give an example of something that is legal but unethical.

11.9 Give an example of something that is illegal but ethical.

11.10 Pick a nanoproduct of your choice and comment on what stage this product is right now in the context of the law of diminishing returns.

Research Assignment

R11.1 Nanotechnology has had a huge impact on the agriculture industry. Consider the use of pesticides and herbicides by engineering targeted delivery, as well as targeted genetic engineering. The role of sensors in soil conditioning is also an important area. Furthermore, nanomaterials are being explored for optimum nutrient delivery, and mineral and vitamin fortification. Pick a topic of your choice about how nanotechnology is impacting agriculture industry, and write a one-page summary.

Reference

1. Landsiedel, Nanomaterials, Wiley-VCH, Bonn, pp. 43–38

Index

© Springer Nature Switzerland AG 2019   109
H. Raza, *Freshman Lectures on Nanotechnology*, Undergraduate Lecture Notes in Physics,
https://doi.org/10.1007/978-3-030-11733-7

Printed in the United States
By Bookmasters